C. Gomez-Reino
M. V. Perez
C. Bao

Gradient-Index Optics

Springer

Berlin
Heidelberg
New York
Barcelona
Hong Kong
London
Milan
Paris
Tokyo

C. Gomez-Reino
M. V. Perez
C. Bao

Gradient-Index Optics

Fundamentals and Applications

Springer

Professor Carlos Gomez-Reino
Professor Maria Victoria Perez
Dr. Carmen Bao
University of Santiago de Compostela
Optics Laboratory, Applied Physics Dept.,
Optics and Optometry School and Faculty of Physics
E - 15782 Santiago de Compostela
Spain
e-mail: facgrc@usc.es

ISBN 978-3-642-07568-1

Library of Congress Cataloging-in-Publication-Data
Die Deutsche Bibliothek – CIP-Einheitsaufnahme

Gomez-Reino, Carlos:
Gradient index optics: fundamentals and applications / C. Gomez-Reino; M. V. Perez; C. Bao. – Berlin; Heidelberg; New York; Barcelona; Hong Kong; London; Milan; Paris; Tokyo: Springer 2002

Springer-Verlag Berlin Heidelberg New York
a member of BertelsmannSpringer Science+Business Media GmbH

http://www.springer.de

Printed in Germany

Cover Design: medio Technologies AG, Berlin
Printed on acid free paper

To Pablo, Clara, Marta, Maria, Javier,
to our families,
and, in memoriam,
to A. Durán

Preface

Currently the terms gradient index or graded index (GRIN) are often used to describe an inhomogeneous medium in which the refractive index varies from point to point [P.1–P.2]. There are three basic gradient index types that have been studied. The first one is the axial gradient. In this case, the refractive index varies in a continuous way along the optical axis of the inhomogeneous medium. The isoindicial surfaces (surfaces of constant index) are planes perpendicular to the optical axis. The second one is the radial GRIN medium in which the index profile varies continuously from the optical axis to the periphery along the transverse direction in such a way that isoindicial surfaces are concentric cylinders about the optical axis. In the special case of the index varying quadratically with the distance in the transverse direction, the medium is referred to as lens-like or self-focusing (selfoc). The last type is the spherical gradient. Spherical gradient may be thought of as a symmetric index around a point so that isoindicial surfaces are concentric spheres. Equations governing propagation of rays through a spherical gradient are similar to those of a geodesic lens.

GRIN media occur commonly in nature. Examples are the crystalline lens and the retinal receptors of the human eye, and the atmosphere of the earth. Shell and continuous refractive index models are used for the crystalline lens [P.3–P.4], and a waveguide model is considered for human photoreceptors [P.5]. The atmosphere of the earth has a refractive index that decreases with height because the density decreases at higher altitudes. Many unusual atmospheric phenomena, a mirage or fata morgana being the best–known example, result from the bending of the rays of light by this gradient [P.6].

In a GRIN medium, the optical rays follow curved trajectories. By an appropriate choice of the refractive index profile, a GRIN medium can have the same effect on light rays as a conventional optical component, such as a prism or a lens. The possibility of using GRIN media in optical systems has been considered for many years, but the manufacture of materials has been the limiting factor in implementing GRIN optical elements until the 1970s. In the last 30 years, however, many different gradient index materials have been manufactured. The revival of GRIN optics has not been casual; it is connected to a considerable degree with the enormous development of optical communications systems, integrated optics, and micro–optics.

Various methods of producing GRIN materials have been developed, but these processes are limited by the small variation of the index, the small depth of the gradient region, and the minimal control over the shape of the resultant index profile. Ion diffusion [P.7] is the most widely used technique for fabrication of glass GRIN materials. They are manufactured by immersing glass elements in a molten salt bath for many hours, during which ion diffusion–exchange occurs through the glass surface and results in the GRIN. Several other methods have been reported. The ion-stuffing method offers the possibility of diffusing large molecules or ions into the glass [P.8], and the sol-gel method involves the synthesis of a multicomponent alkoxide gel that is shaped by a mold [P.9]. Plastic GRIN materials have been manufactured by copolymerization and monomer diffusion [P.10]. Likewise, of particular interest is use of chemical vapor deposition techniques to obtain small–diameter GRIN fibers for optical communications [P.11]. GRIN glasses with large variation of the index and large size have been fabricated by fusing together thin layers of glasses of progressively different refractive indices [P.12].

Historically, Maxwell [P.13] was among the first to consider inhomogeneous media in optics, when, in 1854, he described a GRIN lens, known as Maxwell's fish eye, of spherical symmetry with the property that points on the surface, and within the lens are sharply focused at conjugate points. In 1905, Wood [P.14] constructed a cylindrical lens by a dipping technique whereby a cylinder of gelatin is produced with refractive index axial symmetry. Thin transverse sections with flat end–faces act like converging or diverging lenses, depending on whether the index is a decreasing or increasing function of the radial distance.

Nearly 40 years later, Luneburg [P.1] analyzed ray propagation through inhomogeneous media. He described a variable index, spherically symmetric refracting structure performing perfect geometric imaging between two given concentric spheres. The refractive index profile of such a structure has been expressed by an integral equation. Luneburg solved it explicitly for the case where one of the spheres is of infinite radius, and the second is coincident with the edge of the lens. In the classical Luneburg lens, any parallel bundle of rays passing through the lens converges at a point located on the surface of the lens. Generalizations of the classical Luneburg lens were introduced later. In 1955, Stettler [P.15] considered the cases when the rays completely encircle the center of the lens, and in 1958 Morgan [P.16] allowed refractive index discontinuities. More recently, interesting contributions of the generalized Luneburg lens have been reported by Sochacki [P.17–P.18]. The Luneburg lens is of importance for applications in optics and, for instance, it may be used in integrated optics [P.19–P.24].

Luneburg also analyzed light propagation through a GRIN medium with hyperbolic secant refractive index profile by conformal mapping of a sphere onto a plane. In 1951 and 1954, Mikaelian [P.25] and Fletcher et al. [P.26] describe a radial index profile in terms of a hyperbolic secant function in a cylindrical rod to provide focusing. For this profile, all meridional rays are sharply imaged periodically within the rod. More recently, light propagation in optical fibers [P.27], planar waveguides [P.28], and aspherical laser resonators [P.29] with hyperbolic profiles has been studied as an analogy with the eigenstates of the stationary Schrödinger equation in a hyperbolic potential, which is a special case of the Pöschl–Teller potential [P.30].

In the field of image formation by GRIN lenses, considerable effort has been directed toward geometrical and wave optics. Several methods were used to compute ray trajectories in inhomogeneous media. A convenient way of solving the ray equation was suggested by Montagnino [P.31]. Kapron [P.32] analyzed paraxial ray tracing to derive imaging properties by a GRIN lens. Sands [P.33–P.35] and Moore [P.36–P.37] have used Buchdahl theory for third-order aberrations in rotationally symmetric systems. Buchdahl theory has been extended to fifth-order aberrations by Gupta et al. [P.39]. A study of aberrations of selfoc fibers was made by Brushon [P.40]. Equations for the limiting rays in GRIN lenses were derived by Harrigan [P. 41]. Works on these and another geometrical optics topics have been summarized in the books of Marchand [P.2], Sodha and Ghatak [P.42], Kravtsov and Orlov [P.43], and Greisukh et al. [P.44].

Geometrical optics provide enough knowledge of position, size, and aberrations of the image to allow the design of a GRIN element with reasonably good performance. However, it is also necessary to analyze the performance of the GRIN element by physical optics and guided–wave theory. In this way, Yariv [P.45] studied modal propagation in quadratic index fibers and applied it to the problem of image transmission. Iga et al. [P.46] described imaging and transforming properties in distributed-index lenses and planar waveguides. Gomez-Reino et al. [P.47–P.49] have analyzed paraxial imaging, Fourier transforming transmission, and modal propagation in a GRIN rod, and have also shown that the rod can be represented by a transmittance function equivalent to the conventional lens transmittance function.

Many applications have been found for GRIN materials in science and technology. Linear arrays of radial GRIN rod lenses can be used in the photocopying industry [P.50–P.53]. GRIN lenses can be incorporated within an image intensifier as an image–enhancing system for low light–level applications [P.54] and are used in medical endoscopes with high numerical aperture and low f–number [P.55–P.57]. Light pulses are focused by GRIN lenses onto optical memory and compact disk systems for writing and reading information [P.58–P.61]. Within the communication area, most optical devices for manipulating and processing signals in optical fiber transmission systems include GRIN lenses for carrying out typical functions such as on-axis imaging, collimation, focusing, and off–axis imaging [P.62–P.63]. Taking advantage of these inherent functions of the GRIN lenses, a wide variety of devices have been designed and fabricated. An optical fiber connector is the simplest application of the GRIN lenses for numerical aperture conversion. GRIN attenuators are mainly used to equalize optical signals. Directional couplers and wavelength–division de/multiplexers using GRIN lenses have been developed by inserting beam splitters and interference filters or gratings. One of the key devices in optical transmission systems is the switch and the most typical type is the device where a GRIN lens is moved to switch an optical path. An isolator using GRIN lenses is important to prevent reflected beams from reaching laser sources. An optical bus interconnection system consisting of cascade arrays of GRIN lenses is used as a free–space three–dimensional optical interconnect [P.64]. Finally, GRIN lenses are also used for optical sensing. Intensity–modulated fiber–optic sensors employ

GRIN lenses to improve coupling efficiency and sensing characteristics [P.65–P.67].

This book provides an in–depth, self–contained treatment of GRIN optics and describes the light propagation through inhomogeneous media, which covers basic as well as specialized results. This book should be useful to students, professors, research engineers, physicists, and optometrists in the field of lightwave communications, imaging and transforming systems, and vision sciences as a textbook or a reference book. Chapter 1 deals with the fundamentals of light propagation in inhomogeneous media in the framework of the geometrical optics limit for very large wavenumbers. Chapter 2 provides a general description of light propagation in GRIN media by means of a linear integral transform. Chapters 3 and 4 study the laws of transformation of uniform and non–uniform beams through and by GRIN lenses. Effects of gain and losses in GRIN materials on light propagation are treated in Chap. 5. Chapter 6 concerns GRIN media with hyperbolic secant refractive index profiles. Chapter 7 generalizes the self–imaging phenomenon to GRIN media. Light propagation in the crystalline lens of the human eye by GRIN optics is considered in Chap. 8; Chap. 9 shows some devices that can be made by GRIN lenses.

The authors wish to thank those persons who have contributed to making this book a reality. In particular, the authors are grateful to Prof. L. Garner and Dr. C.R. Fernandez–Pousa for helpful discussions and valuable criticism. The authors also express grateful thanks to M.T. Flores–Arias, M.A. Rama Varela and J.M. Rivas–Moscoso for constructive suggestions.

The writing of this book constitutes an activity in the Education Program of the University of Santiago de Compostela (USC). A short course based on some chapters of this book was offered to graduate students at USC and was well received. The financial support of the Xunta de Galicia and Spanish Government, under PGIDT and TIC plans, for the original work by the authors reported in this book is gratefully acknowledged.

Contents

1 Light Propagation in GRIN Media

1.1 Introduction

Most optical phenomena are described by electromagnetic vectors satisfying the wave equations (partial differential equations). We are especially interested in the electromagnetic vectors for very short wavelengths (very large wavenumbers), i.e., the field that belongs to the realm of geometrical optics. In inhomogeneous media, the vector wave equations for a pair of vectors $\vec{E}$, $\vec{H}$ can be replaced by a simpler scalar wave equation (the Helmholtz equation), provided the refractive index changes very slightly over distances comparable with the wavelength. The Helmholtz equation reduces to a parabolic wave equation for waves propagating approximately along the optical axis of inhomogeneous media.

In this way the chapter considers in succession the vector wave equations, the Helmholtz and parabolic wave equations in the framework of the geometrical optics limit for very large wavenumbers, weakly inhomogeneous media, and propagation of waves that are locally planes. Likewise, we will study light propagation through inhomogeneous media by axial and field rays, two linearly independent particular solutions of the paraxial ray equation.

1.2 Vector Wave Equations

We start with the well-known vector wave equations for the electric intensity vector $\vec{E}(\vec{r})$ and the magnetic intensity vector $\vec{H}(\vec{r})$ of a monochromatic field of frequency ω (periodic field) in isotropic, inhomogeneous, dielectric, nonmagnetic, nondispersive, and linear media

$$\nabla^2\vec{E} + k^2n^2\vec{E} + \vec{\nabla}\left(\frac{\vec{\nabla}n^2}{n^2}\cdot\vec{E}\right) = 0 \tag{1.1a}$$

$$\nabla^2\vec{H} + k^2n^2\vec{H} + \frac{\vec{\nabla}n^2}{n^2}\times\left(\vec{\nabla}\times\vec{H}\right) = 0 \tag{1.1b}$$

where $\nabla^2, \vec{\nabla}$, and $\vec{\nabla}\times$ denote Laplacian, gradient and rotational operators, respectively, n^2 the dimensionless dielectric constant (or dimensionless electric permittivity) and k the wavenumber in vacuum

$$k = \frac{\omega}{c} = \frac{2\pi}{\lambda_0} \tag{1.2}$$

where λ_0 and c are the wavelength and the velocity of light in vacuum.

It may be shown that solutions of Eqs. (1.1) can be expressed by power series in k of the form [1.1, 1.2]

$$\vec{E}(\vec{r}) = \exp\{ikS(\vec{r})\}\sum_{m=0}^{\infty}\frac{\vec{E}_m(\vec{r})}{(ik)^m} \tag{1.3a}$$

$$\vec{H}(\vec{r}) = \exp\{ikS(\vec{r})\}\sum_{m=0}^{\infty}\frac{\vec{H}_m(\vec{r})}{(ik)^m} \tag{1.3b}$$

where $\vec{E}_m$ and $\vec{H}_m$ are complex vector functions of position, and S is "the optical path," a real scalar function of position. The field vectors therefore are developed into asymptotic series and with (1.3) as trial solutions; the vector wave equations lead to a set of relations between $\vec{E}_m$ and $\vec{H}_m$ and S

$$\left(\vec{\nabla}S\right)^2 - n^2 = 0 \tag{1.4a}$$

$$\left(\nabla^2 S + 2\vec{\nabla}S\cdot\vec{\nabla}\right)\vec{E}_0 + \left(\frac{\vec{\nabla}n^2}{n^2}\cdot\vec{E}_0\right)\vec{\nabla}S = 0 \tag{1.4b}$$

$$\left(\nabla^2 S + 2\vec{\nabla}S\cdot\vec{\nabla}\right)\vec{H}_0 + \left(\frac{\vec{\nabla}n^2}{n^2}\times\vec{\nabla}S\right)\times\vec{H}_0 = 0 \tag{1.4c}$$

for m = 0, and

$$\left(\nabla^2 S + 2\vec{\nabla}S\cdot\vec{\nabla}\right)\vec{E}_m + \left(\frac{\vec{\nabla}n^2}{n^2}\cdot\vec{E}_m\right)\vec{\nabla}S + \nabla^2\vec{E}_{m-1} + \vec{\nabla}\left(\frac{\vec{\nabla}n^2}{n^2}\cdot\vec{E}_{m-1}\right) = 0 \tag{1.5a}$$

$$\left(\nabla^2 S + 2\vec{\nabla}S\cdot\vec{\nabla}\right)\vec{H}_m + \left(\frac{\vec{\nabla}n^2}{n^2}\times\vec{\nabla}S\right)\times\vec{H}_m + \nabla^2\vec{H}_{m-1} + \frac{\vec{\nabla}n^2}{n^2}\times\left(\vec{\nabla}\times\vec{H}_{m-1}\right) = 0 \tag{1.5b}$$

for $m \geq 1$.

We are interested in the vectors $\vec{E}_0$ and $\vec{H}_0$ since they determine the electromagnetic field for very large wavenumbers. This limit is the realm of geometrical optics whose basic equation is (1.4a), known as the eikonal equation. The function S is called the eikonal, so that surfaces where S=constant represent surfaces of equal phase or geometrical wavefronts. The surfaces of equal phase define the shape of the electromagnetic field so that the eikonal equation determines the wave propagation in the geometrical optics approximation. $\vec{E}_0$ and $\vec{H}_0$ may be called the geometrical amplitude vectors of the electromagnetic wave. They are perpendicular to each other and tangential to the wavefronts

$$\vec{E}_0 \cdot \vec{\nabla} S = \vec{H}_0 \cdot \vec{\nabla} S = \vec{E}_0^* \cdot \vec{H}_0 = \vec{E}_0 \cdot \vec{H}_0^* = 0 \tag{1.6}$$

Furthermore, the geometrical light rays may be defined as the orthogonal trajectories to the geometrical wavefronts. If $\vec{r}(s)$ denotes the position vector of a point on a ray as a function of s that measures the length along the trajectories, the equation of the rays is written as

$$n\frac{d\vec{r}}{ds} = n\vec{t} = \vec{\nabla} S \tag{1.7a}$$

that is,

$$\frac{dS}{ds} = \vec{t}\vec{\nabla} S = n \tag{1.7b}$$

from Eq. (1.4a); where n is the refractive index of an inhomogeneous medium, and $\vec{t}$ is a unit vector orthogonal to the geometrical wavefront (Fig. 1.1).

Equation (1.7a) specifies the rays by means of the eikonal. If S is known, the trajectories can readily be constructed, since the normal to the geometrical wavefronts at a position $\vec{r}$ is in the direction of the gradient vector of S. Likewise, from Eq. (1.7b) it follows that the arc length ds along a ray between two points on neighboring wavefronts is inversely proportional to the refractive index. Then, integrating Eq. (1.7b) along the ray between these two points P_0 and P_1 gives

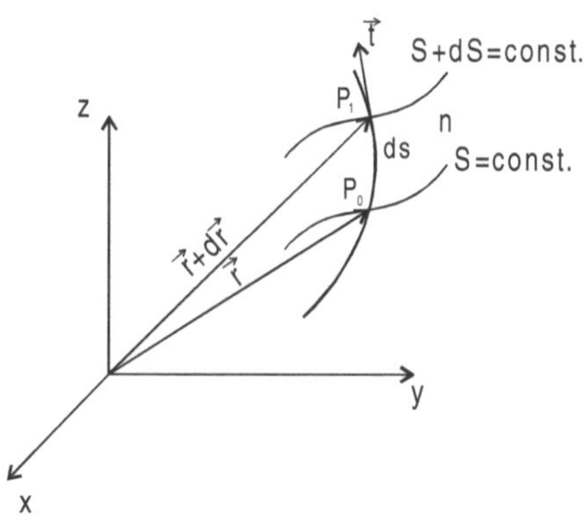

Fig. 1.1. The unit vector orthogonal to the geometrical wavefront and the optical path length.

$$S(P_0,P_1)=\int_{P_0}^{P_1} n\,ds \tag{1.8}$$

which means that the eikonal and, therefore, the phase of the geometrical electromagnetic field undergo changes given by the optical path length of the ray between two points. In this way, the eikonal function may be considered as the characteristic function of the medium, introduced in optics by Hamilton.

Finally, it also follows that when the wavenumber is sufficiently large, the amplitude vectors and the eikonal are connected by Eqs. (1.4b–c) which can be rewritten in compact vectorial form as

$$\nabla^2 S\begin{pmatrix}\vec{E}_0\\ \vec{H}_0\end{pmatrix}+2\vec{\nabla}S\cdot\vec{\nabla}\begin{pmatrix}\vec{E}_0\\ \vec{H}_0\end{pmatrix}+\left[\frac{\vec{\nabla}n^2}{n^2}\cdot\begin{pmatrix}\vec{E}_0\\ \vec{H}_0\end{pmatrix}\right]\vec{\nabla}S=0 \tag{1.9}$$

where Eq. (1.6) has been used. Equation (1.9) is the energy transport equation for geometrical amplitude vectors along the light rays.

1.3 Scalar Wave Equation

The vector wave equations have a rather complicated form. Fortunately, for most problems of light propagation in inhomogeneous media, they can be replaced by simpler wave equations for each component of the electric and magnetic field vectors.

The two first terms in Eqs. (1.1) are of equal order of magnitude, however, it is necessary to estimate the order of magnitude of the third term and to investigate under what conditions it may be neglected. Then, replacing the Laplacian operator by a derivative with respect to some direction b in space, we can write to the order of magnitude of the third term in Eq. (1.1a) that [1.3]

$$\begin{aligned}\vec{\nabla}\left(\frac{\vec{\nabla}n^2}{n^2}\cdot\vec{E}\right)&=\vec{\nabla}\left[\vec{\nabla}(\ln n^2)\cdot\vec{E}\right]\approx\frac{\partial}{\partial b}\left[\frac{\partial}{\partial b}(\ln n^2)\cdot\vec{E}\right]=\frac{\partial(\ln n^2)}{\partial b}\cdot\frac{\partial\vec{E}}{\partial b}+\vec{E}\cdot\frac{\partial^2(\ln n^2)}{\partial b^2}\\ &\approx kn\vec{E}\frac{\partial(\ln n^2)}{\partial b}+\vec{E}\frac{\partial^2(\ln n^2)}{\partial b^2}\end{aligned} \tag{1.10}$$

Note that the second term of Eq. (1.10) is smaller than the first for sufficiently large values of k. By comparing the orders of magnitude of the first term in Eq. (1.10) and the second term of Eq. (1.1a), for instance, we obtain

$$\frac{kn\vec{E}\dfrac{\partial(\ln n^2)}{\partial b}}{k^2n^2\vec{E}}\approx\frac{\lambda}{2\pi}\frac{\partial(\ln n^2)}{\partial b}\approx\frac{\lambda}{2\pi}\frac{\Delta n^2}{n^2\Delta b} \tag{1.11}$$

where λ is the wavelength in inhomogeneous media and Δn^2 represents the variation of n^2 between two close points separated by the distance Δb.

For distance of the order of wavelength ($\Delta b \approx \lambda$), Eq. (1.11) becomes

$$\frac{kn\vec{E}\dfrac{\partial(\ln n^2)}{\partial b}}{k^2 n^2 \vec{E}} \approx \frac{\Delta n^2}{2\pi n^2} \tag{1.12}$$

If we want to neglect the third term in Eq. (1.1a), we must require that the relative change of dielectric constant n^2 over the distance of the order of wavelength must be much smaller than unity, that is

$$\frac{\Delta n^2}{n^2} \langle\langle 1 \tag{1.13}$$

A similar treatment can be made in Eq. (1.1b) for the magnetic field vector since, in this case, we can write to the order of magnitude of the third term that

$$\begin{aligned} \vec{\nabla}(\ln n^2)\times(\vec{\nabla}\times\vec{H}) &= \vec{\nabla}[\vec{\nabla}(\ln n^2)\cdot\vec{H}] - \vec{H}\nabla^2(\ln n^2) \\ &\approx \frac{\partial}{\partial b}\left[\frac{\partial(\ln n^2)}{\partial b}\cdot\vec{H}\right] - \vec{H}\frac{\partial^2(\ln n^2)}{\partial b^2} \approx kn\vec{H}\frac{\partial(\ln n^2)}{\partial b} \end{aligned} \tag{1.14}$$

We conclude that under the condition (1.13) that specifies weak inhomogeneity, the scalar wave equation or the scalar Helmholtz equation

$$\nabla^2\psi + k^2 n^2 \psi = 0 \tag{1.15}$$

is approximately satisfied by each component $\psi(\vec{r})$ of the electric and magnetic field vectors. In other words, the components of the electromagnetic field are uncoupled, and the polarization is preserved through weakly inhomogeneous media (hereafter referred to as gradient-index media or GRIN media). The propagation of light in this kind of medium can therefore be studied by solving the scalar equation.

On the other hand, ψ can also be developed into asymptotic series in k as

$$\psi(\vec{r}) = \exp\{ikS(\vec{r})\}\sum_{m=0}^{\infty}\frac{\psi_m(\vec{r})}{(ik)^m} \tag{1.16}$$

where $\psi_m(\vec{r})$ are real scalar functions of positions.

A set of relations between ψ_m and S may be derived by inserting Eq. (1.16) into Eq. (1.15). One obtains, in particular, for the first term (m=0) of the expansion that

$$\nabla^2\psi_0 + ik[\nabla^2 S + 2\vec{\nabla}S\cdot\vec{\nabla}]\psi_0 + k^2[n^2 - (\vec{\nabla}S)^2]\psi_0 = 0 \tag{1.17}$$

Since ψ_0 and S are assumed to be real and independent of k, the real and imaginary parts of Eq. (1.17) must both vanish. Likewise, for very large k the first term is much smaller than the last term and may, in general, be neglected. Thus,

$$\left(\vec{\nabla}S\right)^2 - n^2 = 0 \tag{1.18a}$$

$$\left(\nabla^2 S + 2\vec{\nabla}S \cdot \vec{\nabla}\right)\psi_0 = 0 \tag{1.18b}$$

Geometrical optics corresponds to this leading term of the expansion. Equation (1.18a) is again the eikonal equation, and (1.18b) is the energy transport equation for the geometrical scalar amplitude ψ_0 along the ray when k is sufficiently large.

Amplitude ψ_0 can be evaluated if we introduce the operator

$$n\frac{d}{ds} = \vec{\nabla}S \cdot \vec{\nabla} \tag{1.19}$$

then, Eq. (1.18b) is written as

$$\frac{1}{\psi_0}\frac{d\psi_0}{ds} = \frac{d(\ln\psi_0)}{ds} = -\frac{1}{2}\frac{\nabla^2 S}{n} \tag{1.20}$$

which is readily integrated to give

$$\psi_0(s) = \psi_0(s_0)\exp\left\{-\frac{1}{2}\int_{s_0}^{s}\frac{\nabla^2 S}{n}ds'\right\} \tag{1.21}$$

Moreover, substituting Eq. (1.21) into (1.16) for m = 0, ψ can be expressed as

$$\psi(s) = \psi_0(s_0)\exp\left\{-\frac{1}{2}\int_{s_0}^{s}\frac{\nabla^2 S}{n}ds'\right\}\exp\{ikS\} \tag{1.22}$$

which represents the geometrical scalar field.

1.4 Parabolic Wave Equation

We now make an approximation applicable to waves that propagate approximately along the z–axis. This approximation reduces the (elliptic) Helmholtz equation to a (parabolic) wave equation.

We assume, without loss of generality, that the dielectric constant of a GRIN medium can be expressed as

$$n^2(x,y,z) = n_0^2(z) + \Delta n^2(x,y,z) \tag{1.23}$$

so that $\Delta n^2/n^2 << 1$, where n_0 is an arbitrary function of z that represents the refractive index along the z-axis.

For waves propagating in a direction close to the z-axis, the scalar field ψ can be written as

$$\psi(\vec{r}) = \phi(\vec{r}) \exp\left\{ ik \int_{z_0}^{z} n_0(z') dz' \right\} \tag{1.24}$$

Equation (1.24) assumes that the variation of ϕ with position must be slow within the distance of a wavelength in the z–direction, so that the propagation direction of the wave bends slightly with respect to z, and the field approximately maintains plane wave nature. Inserting Eq. (1.24) into Eq. (1.15) and neglecting the second derivative of ϕ with respect to z, we obtain

$$\nabla_\perp^2 \phi + 2ikn_0 \frac{\partial \phi}{\partial z} + ik\phi \frac{dn_0}{dz} + k^2 \left(n^2 - n_0^2\right)\phi = 0 \tag{1.25}$$

where $\nabla_\perp^2$ is the transverse Laplacian operator

$$\nabla_\perp^2 = \nabla^2 - \frac{\partial^2}{\partial z^2} \tag{1.26}$$

If we introduce the changes of field and variable

$$\theta(\vec{r}) = n_0^{1/2}(z)\phi(\vec{r}) \tag{1.27a}$$

$$\eta = \int_{z_0}^{z} \frac{dz'}{n_0(z')} \tag{1.27b}$$

Equation (1.25) becomes

$$\nabla_\perp^2 \theta + 2ik \frac{\partial \theta}{\partial \eta} + k^2 \left(n^2 - n_0^2\right)\theta = 0 \tag{1.28}$$

Equation (1.28) represents a parabolic wave equation that has the form of a non–stationary Schrödinger equation, where η represents time and n represents potential. It can be rewritten in the operator formulation as

$$2ik \frac{\partial \theta}{\partial \eta} = H\theta \tag{1.29}$$

where H is the Hamiltonian operator given by

$$H = -\nabla_\perp^2 - k^2 \left(n^2 - n_0^2\right) \tag{1.30}$$

Equations (1.28–1.30) can be simplified further, if we consider that for weakly inhomogeneous media

$$n^2 - n_0^2 \approx 2n_0\Delta n \tag{1.31}$$

that is,

$$H \approx -\nabla_\perp^2 - 2k^2 n_0 \Delta n \tag{1.32}$$

where $\Delta n = n - n_0$.

Likewise, ψ can be developed in series as before, and the geometrical field will be given by the first term of the expansion [1.4]

$$\psi(\vec{r}) = \psi_0^P(\vec{r})\exp\{ikS^P(\vec{r})\} \tag{1.33}$$

where ψ_0^P and S^P are the parabolic geometrical amplitude and the parabolic eikonal function, respectively. They are real functions of position that are slowly varying in comparison with the wavelength in the z–direction, so that $\partial^2 S^P/\partial z^2$ and $\partial S^P/\partial z \cdot \partial \psi_0^P/\partial z$ can be neglected.

Then, Eqs. (1.18a–b) are now expressed as

$$\left(\vec{\nabla} S^P\right)^2 - n^2 = 0 \tag{1.34a}$$

$$\psi_0^P \nabla_\perp^2 S^P + 2\vec{\nabla}_\perp S^P \cdot \vec{\nabla}_\perp \psi_0^P = 0 \tag{1.34b}$$

where $\vec{\nabla}_\perp$ denotes the transverse gradient operator.

From Eq. (1.34a) it follows that

$$\frac{\partial S^P}{\partial z} = \left[n^2 - \left(\frac{\partial S^P}{\partial x}\right)^2 - \left(\frac{\partial S^P}{\partial y}\right)^2\right]^{1/2} \tag{1.35}$$

For waves propagating close to the z-axis, Eq. (1.35) can be approximated as

$$\frac{\partial S^P}{\partial z} \approx n - \frac{1}{2n}\left[\left(\frac{\partial S^P}{\partial x}\right)^2 + \left(\frac{\partial S^P}{\partial y}\right)^2\right] \approx n - \frac{1}{2n_0}\left(\vec{\nabla}_\perp S^P\right)^2 \tag{1.36}$$

where the terms of order higher than second in $\partial S^P/\partial x$ and $\partial S^P/\partial y$ have been neglected. We have also assumed that the transverse variation of n is small so that n can be replaced by n_0 in the second term of Eq. (1.36). Then, the parabolic eikonal function satisfies approximately the following equation

$$\frac{1}{2}\left(\vec{\nabla}_\perp S^P\right)^2 + n_0 \frac{\partial S^P}{\partial z} = n n_0 \tag{1.37}$$

On the other hand, integrating Eq. (1.34b), the geometrical field may be written as

$$\psi_0(\vec{r}) = \psi_0^P(\vec{r}_0)\exp\left\{-\frac{1}{2}\int_{z_0}^{z} \frac{\nabla_\perp^2 S^P}{n_0} dz'\right\}\exp\{ikS^P\} \tag{1.38}$$

where Eq. (1.33) and

$$\vec{\nabla}_{\perp} S_{p} \cdot \vec{\nabla}_{\perp} = n_{0} \frac{d}{dz} \tag{1.39}$$

have been used.

Equations (1.22) and (1.38) are elliptic and parabolic leading solutions for the scalar field, respectively, and they represent the propagation of the geometrical complex amplitude between two arbitrary points joined by all possible rays. They become the kernel functions of the integral equation used to find the complex amplitude distribution on an arbitrary surface if the scalar field at the initial surface is known in the elliptic and parabolic approximations.

1.5 Ray Optics: Axial and Field Rays

For inhomogeneous media the eikonal function (optical path length) given by Eq. (1.8) can be written as

$$S = \int_{P_0}^{P_1} L(x, y, z, x', y', z'; s) ds \tag{1.40}$$

where ds is a differential arc length along a ray trajectory between P_0 and P_1

$$ds = \sqrt{dx^2 + dy^2 + dz^2} \tag{1.41}$$

L is the optical Lagrangian [1.5]

$$L = n(x, y, z)\left(x'^2 + y'^2 + z'^2\right)^{1/2} \tag{1.42}$$

and the prime denotes derivative with respect to s.

It is quite customary to deduce the ray equation from Fermat's principle, which asserts that "the light ray between two points is the curve for which the optical path attains an extreme value." We now briefly summarize the steps to obtain the ray equation as an extremum condition. Fermat's principle assumes that

$$S = \int_{P_0}^{P_1} L ds = \text{extremum} \tag{1.43a}$$

Equation (1.43a) is of the same form as Hamilton's principle in classical mechanics where s denotes time.

The extremum condition of S is expressed as

$$\delta S = 0 \tag{1.43b}$$

The solution of the variational problem (1.43b) is given by Euler's equations

$$\frac{d}{ds}\left(\frac{\partial L}{\partial x'}\right)=\frac{\partial L}{\partial x} \tag{1.44a}$$

$$\frac{d}{ds}\left(\frac{\partial L}{\partial y'}\right)=\frac{\partial L}{\partial y} \tag{1.44b}$$

$$\frac{d}{ds}\left(\frac{\partial L}{\partial z'}\right)=\frac{\partial L}{\partial z} \tag{1.44c}$$

The vectorial form of these three second-order equations is

$$\frac{d}{ds}\left(n\frac{d\vec{r}}{ds}\right)=\vec{\nabla}n \tag{1.45}$$

where $\vec{\nabla}n$ is the gradient of n and Eq. (1.42) has been used.

Equation (1.45) is the ray equation and specifies the light ray directly in terms of the refractive index. The ray trajectory will be a continuous curve between P_0 and P_1, which will also have a continuous tangent described by the position vector $\vec{r}(s)$. The ray equation can be derived in other ways by differentiating, for instance, Eq. (1.7a) with respect to s.

Equation (1.45) can be rewritten as

$$n'\vec{t}+n\vec{\kappa}=\vec{\nabla}n \tag{1.46}$$

where $\vec{t}={d\vec{r}}/{ds}=\vec{r}\,'$ is the unit tangential vector at a point of the ray and

$$\vec{\kappa}(s)=\vec{r}\,''=|\vec{\kappa}(s)|\vec{\nu} \tag{1.47}$$

is the curvature vector of the ray. The unit principal normal vector at a point of the ray $\vec{\nu}$ is defined as

$$\vec{\nu}=\frac{\vec{r}\,''}{|\vec{r}\,''|}=\frac{\vec{t}\,'}{|\vec{t}\,'|} \tag{1.48}$$

Then, from Eq. (1.46) it follows that the gradient of n lies on the osculating plane of the ray.

Since

$$\vec{t}\cdot\vec{t}\,'=\vec{t}\cdot\vec{\kappa}=0 \tag{1.49}$$

scalar multiplication of (1.46) by $\vec{\kappa}$ gives

$$|\vec{\kappa}|=\vec{\nu}\frac{\vec{\nabla}n}{n}=\vec{\nu}\cdot\vec{\nabla}(\ln n) \tag{1.50}$$

that is, the ray bends toward the region of higher refractive index.

On the other hand, the differential equation of rays may be put into Cartesian coordinates if one variable is taken as an independent variable. The z–coordinate is usually chosen to coincide with the axis of the optical system. Therefore, Euler's equations become

$$\frac{d}{dz}\left(\frac{\partial L}{\partial \dot{x}}\right)=\frac{\partial L}{\partial x}$$
$$\frac{d}{dz}\left(\frac{\partial L}{\partial \dot{y}}\right)=\frac{\partial L}{\partial y} \tag{1.51}$$

where L and S are now given, respectively, by

$$L(x, y, \dot{x}, \dot{y}; z)=n(x, y, z)\left(1+\dot{x}^2+\dot{y}^2\right)^{1/2} \tag{1.52a}$$

$$S=\int_{z_0}^{z_1} L dz \tag{1.52b}$$

where $\dot{x}$ denotes the derivative of x with respect to z.

From Eq. (1.52a) it follows that Eq. (1.51) is expressed as

$$\frac{d}{dz}\left[\frac{n\dot{x}}{\sqrt{1+\dot{x}^2+\dot{y}^2}}\right]=\sqrt{1+\dot{x}^2+\dot{y}^2}\,\frac{\partial n}{\partial x}$$
$$\frac{d}{dz}\left[\frac{n\dot{y}}{\sqrt{1+\dot{x}^2+\dot{y}^2}}\right]=\sqrt{1+\dot{x}^2+\dot{y}^2}\,\frac{\partial n}{\partial y} \tag{1.53}$$

Equations (1.53) describe the light ray by two functions x(z) and y(z) with a continuous derivative with respect to z.

The quantities

$$\cos\alpha=\frac{\dot{x}}{\sqrt{1+\dot{x}^2+\dot{y}^2}}$$
$$\cos\beta=\frac{\dot{y}}{\sqrt{1+\dot{x}^2+\dot{y}^2}} \tag{1.54}$$
$$\cos\gamma=\frac{1}{\sqrt{1+\dot{x}^2+\dot{y}^2}}$$

are the direction cosines of the light with respect to x–, y–, z–axes and

$$p = n\cos\alpha \tag{1.55a}$$
$$q = n\cos\beta \tag{1.55b}$$
$$1 = n\cos\gamma \tag{1.55c}$$

are the corresponding optical direction cosines that are related by

$$n^2 = p^2 + q^2 + l^2 \tag{1.56}$$

The Euler's equations become, with the aid of Eqs. (1.54–1.56)

$$\begin{aligned} \dot{p} &= \frac{n}{l}\frac{\partial n}{\partial x} = \frac{\partial l}{\partial x} \\ \dot{q} &= \frac{n}{l}\frac{\partial n}{\partial y} = \frac{\partial l}{\partial y} \end{aligned} \tag{1.57}$$

Likewise, $\dot{x}$ and $\dot{y}$ in terms of the optical direction cosines can be expressed as

$$\begin{aligned} \dot{x} &= \frac{p}{l} = -\frac{\partial l}{\partial p} \\ \dot{y} &= \frac{q}{l} = -\frac{\partial l}{\partial q} \end{aligned} \tag{1.58}$$

Thus, the position x(z), y(z) and the optical direction cosines p(z), q(z) of the ray satisfy a system of canonical equations given by (1.57–1.58).

Equation (1.53) can also expressed as [1.6]

$$\begin{aligned} n\ddot{x} &= \left(1 + \dot{x}^2 + \dot{y}^2\right)\left(\frac{\partial n}{\partial x} - \dot{x}\frac{\partial n}{\partial z}\right) \\ n\ddot{y} &= \left(1 + \dot{x}^2 + \dot{y}^2\right)\left(\frac{\partial n}{\partial y} - \dot{y}\frac{\partial n}{\partial z}\right) \end{aligned} \tag{1.59}$$

which can be rewritten as

$$\begin{aligned} n\frac{\partial l}{\partial p}\frac{\partial n}{\partial z} + p\dot{l} &= 0 \\ n\frac{\partial l}{\partial q}\frac{\partial n}{\partial z} + q\dot{l} &= 0 \end{aligned} \tag{1.60}$$

where canonical equations were used.

From Eqs. (1.60) it follows that for GRIN media with only transverse variation of the refractive index, l is constant since $\partial n/\partial z = 0$. Therefore the third optical direction cosine is invariant along the ray.

Since, in general, it is not possible to solve the canonical equations exactly, we will derive approximate solutions. The lowest order approximation is called the paraxial approximation. This approximation is determined by the condition that the ray is very close to the z–axis, and also that the optical direction cosines p, q of the ray are small [1.1]. The deviation from this approximation determines the aberrations present in media [1.7].

We shall now be concerned with the paraxial approximation for systems of rotational symmetry. We consider x, y small quantities and develop the dielectric constant around the z axis. Expressing n^2 as a McLaurin series

$$n^2(x,y,z) = n_0^2(z) \pm |n_{1z}^2|(x^2+y^2) \pm \frac{1}{2}|n_{2z}^2|(x^2+y^2)^2 \pm \cdots \tag{1.61}$$

where n_0^2, n_{1z}^2, and n_{2z}^2 are the dielectric constants along the z–axis and the first and the second partial derivatives of n^2 with respect to z evaluated at x=y=0, respectively, it follows that for GRIN media the refractive index is given by

$$n(x,y,z) = n_0(z) \pm \frac{|n_{1z}^2|}{2n_0(z)}(x^2+y^2) \pm \text{higher-order terms} \tag{1.62}$$

It is customary to write the above equations to a first-order approximation as

$$n^2(x,y,z) = n_0^2(z)[1 \pm g^2(z)(x^2+y^2)] \tag{1.63a}$$

$$n(x,y,z) = n_0(z)\left[1 \pm \frac{g^2(z)}{2}(x^2+y^2)\right] \tag{1.63b}$$

where

$$g^2(z) = \frac{|n_{1z}^2(z)|}{n_0^2(z)} \rangle 0 \tag{1.64}$$

is a slowly varying function with z that characterizes the dielectric constant or refractive index distributions.

The transverse parabolic variation of n^2 or n decreases from the axis when the second term in Eq. (1.63) is substracted, and, on the contrary, the transverse parabolic variation increases from the axis when the terms are added.

Likewise, if we also consider p, q small quantities for which the ray bends slightly with respect to the z–axis, Eqs. (1.55a–b) reduce to

$$p = n_0(z)\dot{x} \tag{1.65a}$$

$$q = n_0(z)\dot{y} \tag{1.65b}$$

If the ray lies in the x–z plane, then $\dot{x} \approx \tan\theta$, where θ is the angle made by the tangent to the ray and the z-axis. Then $\dot{x}$, $\dot{y}$ represent the slope of the ray at a certain point P.

On the other hand, Eqs. (1.53) to the paraxial approximation become

$$\dot{p} = \frac{d}{dz}[n_0(z)\dot{x}] = \pm n_0(z)g^2(z)x \tag{1.66a}$$

$$\dot{q} = \frac{d}{dz}[n_0(z)\dot{y}] = \pm n_0(z)g^2(z)y \tag{1.66b}$$

The canonical equations (1.65) and (1.66) are called paraxial equations, and their solutions are paraxial rays that can be considered as approximations of the exact light rays in the neighborhood of the axis. Since our GRIN media are completely symmetric in x, y and p, q, it is enough if we consider only one set of equations, say for x, p, and the equations for y, q follow from analogy.

Equation (1.66a) can be rewritten as the following linear second-order differential equation

$$\ddot{x} + \frac{\dot{n}_0(z)}{n_0(z)}\dot{x} \pm g^2(z)x = 0 \tag{1.67}$$

Solutions of Eq. (1.67) are much easier to obtain if n_0 is constant. In that case we obtain

$$\ddot{x} \pm g^2(z)x = 0 \tag{1.68}$$

It is important to note, however, that with the change of variable given by Eq. (1.27b), Eq. (1.67) can always be transformed into Eq. (1.68) without approximation. Hence there is no loss of generality in using Eq. (1.68) in place of Eq. (1.67).

Likewise, assuming Eqs. (1.63) with the negative sign, Eq. (1.68) possesses solutions with an infinite number of zeros in the interval $(0,\infty)$, that is, Eq. (1.68) has oscillatory solutions. However, when the positive sign is present in Eqs. (1.63), the solutions of the differential equation (1.68) are non–oscillatory [1.8].

We are interested in oscillatory solutions in order to study paraxial imaging and transforming properties through GRIN media. Therefore as mentioned above, the basis ideas are to use Eq. (1.68) in place of Eq. (1.67), and to consider one set of equations for x, p. With this in mind, if we assume a dielectric constant or a refractive index of the form

$$n^2(x,y,z) = n_0^2\left[1 - g^2(z)\left(x^2 + y^2\right)\right] \tag{1.69a}$$

$$n(x,y,z) = n_0\left[1 - \frac{g^2(z)}{2}\left(x^2 + y^2\right)\right] \tag{1.69b}$$

it follows that Eq. (1.65a) and (1.68) can be written as

$$p = n_0\dot{x} \tag{1.70a}$$

$$\ddot{x} + g^2(z)x = 0 \tag{1.70b}$$

The solutions of these equations give the position x(z) and the optical direction cosine p(z) of the paraxial ray at a point P, and they can be expressed as a linear combination of two linearly independent particular solutions.

These particular solutions are the axial and the field rays defined by Luneburg [1.2]. The axial ray is the solution

$$x_a = H_a(z) \tag{1.71a}$$

$$p_a = n_0\dot{H}_a(z) \tag{1.71b}$$

of Eqs. (1.70) that, at an arbitrary reference plane such as z = 0, satisfies the boundary conditions

$$H_a(0)=0 \tag{1.72a}$$

$$p_a(0)=n_0 \Rightarrow \dot{H}_a(0)=1 \tag{1.72b}$$

The field ray is a second solution

$$x_f = H_f(z) \tag{1.73a}$$

$$p_f = n_0\dot{H}_f(z) \tag{1.73b}$$

of Eqs. (1.70) with the boundary conditions

$$H_f(0)=1 \tag{1.74a}$$

$$p_f(0)=0 \Rightarrow \dot{H}_f(0)=0 \tag{1.74b}$$

where $\dot{H}_a, \dot{H}_f$ are the slopes of the axial and field rays, respectively.

Thus the axial ray is a paraxial ray that originates at the point of the axis, and the field ray is a paraxial ray that leaves the reference plane parallel to the axis (Fig. 1.2).

With the aid of these two rays we can express any other paraxial ray in the form

$$x(z)=\alpha H_a(z)+\beta H_f(z) \tag{1.75a}$$

$$p(z)=n_0\dot{x}(z)=\alpha p_a(z)+\beta p_f(z) \tag{1.75b}$$

where α, β are constants that depend on the boundary conditions of the ray.

From Eqs. (1.72) and (1.74), it follows that Eqs. (1.75) become

$$x(z)=H_f(z)x_0+\frac{H_a(z)}{n_0}p_0 \tag{1.76a}$$

$$p(z)=p_f(z)x_0+p_a(z)\frac{p_0}{n_0}=n_0\dot{H}_f(z)x_0+\dot{H}_a(z)p_0 \tag{1.76b}$$

where Eqs. (1.71) and (1.73) have been used.

These equations represent a paraxial ray that leaves the reference plane z = 0 at a point x_0 with the direction p_0.

Eqs. (1.76) can be written for the position and the slope of the ray as

$$x(z)=H_f(z)x_0+H_a(z)\dot{x}_0 \tag{1.77a}$$

$$\dot{x}(z)=\dot{H}_f(z)x_0+\dot{H}_a(z)\dot{x}_0 \tag{1.77b}$$

where $\dot{x}_0$ is the slope of the ray at a position x_0 on the plane z = 0.

Likewise, two paraxial rays x_1, p_1 and x_2, p_2 are related by Lagrange's invariant. At any transverse plane z, it is given by

$$p_1(z)x_2(z)-x_1(z)p_2(z)=\text{constant} \tag{1.78}$$

that is, the Wronskian is a constant.

In the case of the axial and field rays, we find that

$$p_a(z)H_f(z) - H_a(z)p_f(z) = n_0 \tag{1.79a}$$

$$\dot{H}_a(z)H_f(z) - H_a(z)\dot{H}_f(z) = 1 \tag{1.79b}$$

where boundary conditions have been used.

With the aid of this result we can show that it is possible to determine the field ray and thus any other paraxial ray if, for instance, the axial ray is known.

From Eq. (1.79b) it follows that

$$\frac{d}{dz}\left(\frac{H_f}{H_a}\right) = -\frac{1}{H_a^2} \tag{1.80}$$

and hence

$$H_f(z) = -H_a(z)\int_0^z \frac{dz'}{H_a^2(z')} \tag{1.81}$$

From

$$\dot{H}_f(z)\ddot{H}_a(z) - \ddot{H}_f(z)\dot{H}_a(z) = -\frac{\ddot{H}_a(z)}{H_a(z)} = g^2(z) \tag{1.82}$$

where Eqs. (1.70b) and (1.79b) have been used, it follows that

$$\frac{d}{dz}\left(\frac{\dot{H}_f}{\dot{H}_a}\right) = -\frac{g^2}{\dot{H}_a^2} \tag{1.83}$$

and this yields

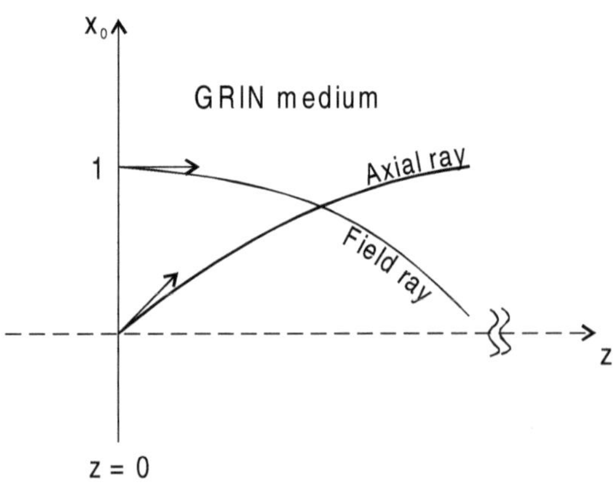

Fig. 1.2. The axial and field rays with initial conditions.

$$\dot{H}_f(z) = -\dot{H}_a(z)\int_0^z \frac{g^2(z')}{\dot{H}_a^2(z')}dz' \tag{1.84}$$

Let us now consider any other paraxial ray. The Lagrange's invariant says

$$x(z)\dot{H}_a(z) - \dot{x}(z)H_a(z) = \text{cons tan t} \tag{1.85}$$

We can find the constant value for z = 0. Then Eq. (1.85) is written as

$$x(z)\dot{H}_a(z) - \dot{x}(z)H_a(z) = x_0 \tag{1.86}$$

and Eq. (1.82) is easily generalized to give

$$\dot{x}(z)\ddot{H}_a(z) - \ddot{x}(z)\dot{H}_a(z) = g^2(z)x_0 \tag{1.87}$$

From Eqs. (1.86–1.87) it follows that

$$x(z) = -x_0 H_a(z)\int_0^z \frac{dz'}{H_a^2(z')} \tag{1.88a}$$

$$\dot{x}(z) = -x_0 \dot{H}_a(z)\int_0^z \frac{g^2(z')}{\dot{H}_a^2(z')}dz' \tag{1.88b}$$

Obviously, the above question can also be solved if the field ray, instead of the axial ray, is known.

The linear algebraic nature of Eqs. (1.76–1.77) is used to generate matrices that describe the paraxial propagation of light through a GRIN medium. The resulting matrix equations are of the form

$$\begin{pmatrix} x(z) \\ p(z) \end{pmatrix} = M'(z)\begin{pmatrix} x_0 \\ p_0 \end{pmatrix} = \begin{pmatrix} H_f(z) & \dfrac{H_a(z)}{n_0} \\ n_0\dot{H}_f(z) & \dot{H}_a(z) \end{pmatrix}\begin{pmatrix} x_0 \\ p_0 \end{pmatrix} \tag{1.89a}$$

$$\begin{pmatrix} x(z) \\ \dot{x}(z) \end{pmatrix} = M(z)\begin{pmatrix} x_0 \\ \dot{x}_0 \end{pmatrix} = \begin{pmatrix} H_f(z) & H_a(z) \\ \dot{H}_f(z) & \dot{H}_a(z) \end{pmatrix}\begin{pmatrix} x_0 \\ \dot{x}_0 \end{pmatrix} \tag{1.89b}$$

such that

$$\det M'(z) = \det M(z) = 1 \tag{1.90}$$

The matrices M and M' whose elements are

$$A = H_f(z);\; B = H_a(z);\; C = \dot{H}_f(z);\; D = \dot{H}_a(z) \tag{1.91a}$$

$$A' = A;\; B' = \frac{B}{n_0};\; C' = n_0 C;\; D' = D \tag{1.91b}$$

characterize the propagation completely, since they permit x, p or x, $\dot{x}$ to be determined for x_0, p_0 or x_0, $\dot{x}_0$. They are known as the ray-transfer matrices or ABCD matrices. Note that for free space $n_0 = 1$ and $M' = M = \begin{pmatrix} 1 & z \\ 0 & 1 \end{pmatrix}$.

On the other hand, we suppose that the axial ray cuts the z–axis at least twice. If z_1 and z_2 are two consecutives zeros of H_a, the field ray H_f has one, and only one, zero in the interval (z_1, z_2), that is, H_f cuts the axis at least once, and the converse is also true. This property, named interlacing of zeros [1.8], will be used to evaluate paraxial imaging and transforming conditions in GRIN media, since H_a and H_f are two linearly independent oscillatory solutions of Eqs. (1.70). In this way, both rays can be expressed as [1.9, 1.10]

$$H_a(z) = [g_0 g(z)]^{-1/2} \sin\left[\int_0^z g(z')dz'\right] \tag{1.92a}$$

$$H_f(z) = \left[\frac{g_0}{g(z)}\right]^{1/2} \cos\left[\int_0^z g(z')dz'\right] \tag{1.92b}$$

where g_0 is the value of g(z) at z = 0.

These equations obey Eq. (1.70b) only approximately. The substitution provides

$$\ddot{g}g - \frac{3}{2}\dot{g}^2 = g^3\left[\frac{g}{2}\left(\frac{\dot{g}}{g^2}\right)^2 + \frac{d}{dz}\left(\frac{\dot{g}}{g^2}\right)\right] = 0 \tag{1.93}$$

Equation (1.93) is, in general, almost zero for small variations of g(z) over a wavelength distance, that is

$$\frac{|\dot{g}(z)|}{g^2(z)} << 1 \tag{1.94}$$

If condition (1.94) is fulfilled, the slope of the axial and field rays can be approximated as

$$\dot{H}_a(z) = \left[\frac{g(z)}{g_0}\right]^{1/2} \cos\left[\int_0^z g(z')dz'\right] \tag{1.95a}$$

$$\dot{H}_f(z) = -[g_0 g(z)]^{1/2} \sin\left[\int_0^z g(z')dz'\right] \tag{1.95b}$$

Then the oscillatory rays preserve the boundary conditions as well as satisfy the Lagrange's invariant.

For GRIN media with only transverse variation of the refractive index in which g(z) = constant = g_0, position and slope of the axial and field rays reduce to

$$H_a(z)=\frac{\sin(g_0 z)}{g_0};\quad H_f(z)=\cos(g_0 z) \tag{1.96a}$$

$$\dot{H}_a(z)=\cos(g_0 z);\quad \dot{H}_f(z)=-g_0\sin(g_0 z) \tag{1.96b}$$

In some cases it is more convenient to choose the axial and field rays in a different manner. One can specify the position of these rays at two parallel reference planes, such as z = 0 and $z = z_1$ [1.2]. The axial ray is defined by the conditions

$$H_a(0)=0 \tag{1.97a}$$

$$H_a(z_1)=1 \tag{1.97b}$$

and the field ray by

$$H_f(0)=1 \tag{1.98a}$$

$$H_f(z_1)=0 \tag{1.98b}$$

The axial ray leaves the axis at z = 0 and cuts the plane z_1 at unit height. Reverse conditions occur for the field ray (Fig. 1.3). Both rays are two linearly independent solutions of the paraxial ray equation with conditions (1.97–1.98), and they are given by

$$H_a(z)=\frac{g(z_1)}{g^{1/2}(z)\left[g^{1/2}(z_1)\exp\left\{i\int_0^{z_1}g(z')dz'\right\}-g_0^{1/2}\exp\left\{-i\int_0^{z_1}g(z')dz'\right\}\right]}$$
$$\cdot\left[\exp\left\{i\int_0^{z}g(z')dz'\right\}-\left[\frac{g_0}{g(z)}\right]^{1/2}\exp\left\{-i\int_0^{z}g(z')dz'\right\}\right] \tag{1.99a}$$

$$H_f(z)=\frac{g_0}{g^{1/2}(z)\left[g_0^{1/2}-g^{1/2}(z_1)\exp\left\{i2\int_0^{z_1}g(z')dz'\right\}\right]} \tag{1.99b}$$
$$\cdot\left[\exp\left\{i\int_0^{z}g(z')dz'\right\}-\left[\frac{g(z_1)}{g(z)}\right]^{1/2}\exp\left\{i2\int_0^{z_1}g(z')dz'\right\}\exp\left\{-i\int_0^{z}g(z')dz'\right\}\right]$$

For GRIN media with only transverse variation of the refractive index, Eqs. (1.99a–b) reduce to

$$H_a(z)=\frac{\sin(g_0 z)}{\sin(g_0 z_1)} \tag{1.99c}$$

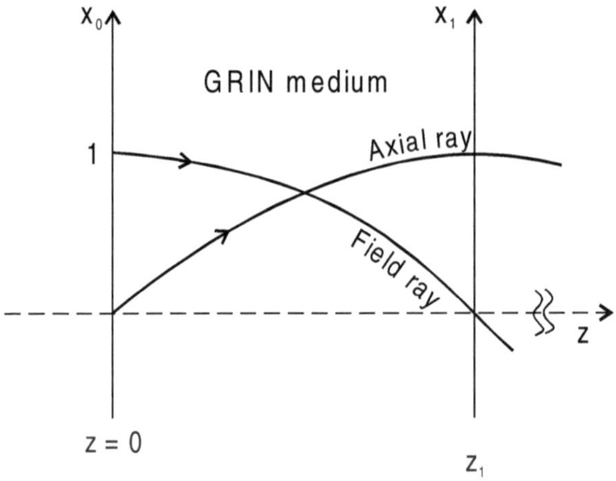

Fig. 1.3. The axial and field rays with boundary conditions.

$$H_f(z) = \cos(g_0 z) - \frac{\sin(g_0 z)}{\tan(g_0 z_1)} \tag{1.99d}$$

Now the position and slope of any ray at z can be written as

$$x(z) = H_f(z)x_0 + H_a(z)x_1 \tag{1.100a}$$

$$\dot{x}(z) = \dot{H}_f(z)x_0 + \dot{H}_a(z)x_1 \tag{1.100b}$$

where $\dot{H}_a$ and $\dot{H}_f$ are the slope of the axial and field rays at z.

Equations (1.100) represent a paraxial ray that originates at a point x_0 of the reference plane $z = 0$ and intersects the other reference plane z_1 at the point x_1.

Finally, we consider the case of a surface S that separates two media of indices of refraction n_1 and n_2. Snell's law of refraction at any point P of the surface can be expressed as

$$n_1(\vec{v} \times \vec{t}_1) = n_2(\vec{v} \times \vec{t}_2) \tag{1.101}$$

where $\vec{v}$ is the unit normal vector at the point P, and $\vec{t}_1, \vec{t}_2$ are unit tangential vectors of the light ray at P on both sides of S (Fig. 1.4). If θ_1 and θ_2 are the angles made by the incident and refracted rays with $\vec{v}$, then Eq. (1.101) yields

$$n_1 \sin\theta_1 = n_2 \sin\theta_2 \tag{1.102}$$

Equation (1.102) for paraxial approximation reduces to

$$n_1\theta_1 = n_2\theta_2 \tag{1.103}$$

With the aid of Eq. (1.103), we apply the paraxial ray equations to a GRIN medium illuminated by an external source. Let z = 0 be a plane that separates the GRIN medium and free space. We assume the reference plane is illuminated by an on-axis point source at a distance d_1 from z = 0 (Fig. 1.5). If we set $n_1 = 1$ and $\theta_1 \approx \frac{x_0}{d_1}$, the slope at a point x_0 on z = 0 is given by

$$\dot{x}_0 = \tan\theta_2 \approx \theta_2 \approx \frac{x_0}{n_0 d_1} \tag{1.104}$$

where n_2, refractive index of GRIN medium, has been replaced by n_0, index along z–axis, since the transverse variation of refractive index is very small for the paraxial approximation in the weakly inhomogeneous medium. Hence Eqs. (1.77) can be rewritten as

$$x(z) = \left[H_f(z) + \frac{H_a(z)}{n_0 d_1}\right] x_0 \tag{1.105a}$$

$$\dot{x}(z) = \left[\dot{H}_f(z) + \frac{\dot{H}_a(z)}{n_0 d_1}\right] x_0 \tag{1.105b}$$

Likewise, taking into account Eqs. (1.92), the ray position given by Eq. (1.105a) can also be expressed as

$$x(z) = \frac{x_0}{[g_0 g(z)]^{1/2}}\left\{g_0 \cos\left[\int_0^z g(z')dz'\right] + \frac{1}{n_0 d_1}\sin\left[\int_0^z g(z')dz'\right]\right\} \tag{1.106}$$

that is

$$x(z) = R(z)\sin\left[\int_0^z g(z')dz' + \delta_0\right] \tag{1.107}$$

where

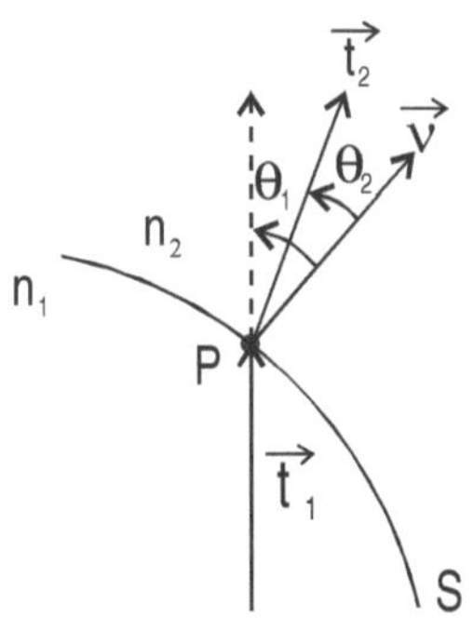

Fig. 1.4. Snell's law of refraction.

$$R(z) = \frac{\sqrt{1+(n_0 g_0 d_1)^2}}{n_0 d_1 [g_0 g(z)]^{1/2}} x_0 = \frac{\sqrt{\dot{x}_0^2 + (g_0 x_0)^2}}{[g_0 g(z)]^{1/2}} \tag{1.108a}$$

$$\delta_0 = \tan^{-1}(n_0 g_0 d_1) = \tan^{-1}\left(\frac{g_0 x_0}{\dot{x}_0}\right) \tag{1.108b}$$

and Eq. (1.104) has been used.

Equation (1.107) shows that the paraxial refracted ray path in the GRIN medium is a sinusoidal path with amplitude R(z) and initial phase δ_0 [1.11].

From Eqs. (1.107) or (1.105b) it also follows that the refracted ray slope has a similar behavior, since

$$\dot{x}(z) = g(z) R(z) \cos\left[\int_0^z g(z')dz' + \delta_0\right] \tag{1.109}$$

Of particular importance are the cases where the source is located at infinite distance or on the input plane. Ray position and ray slope at any plane $z > 0$ when the medium is illuminated by a collimated light beam ($\dot{x}_0 = 0$ or $d_1 \to \infty$) are given by

$$x(z) = H_f(z) x_0 \tag{1.110a}$$

$$\dot{x}(z) = \dot{H}_f(z) x_0 \tag{1.110b}$$

This means that position and slope of the ray are proportional to position and slope of the field ray.

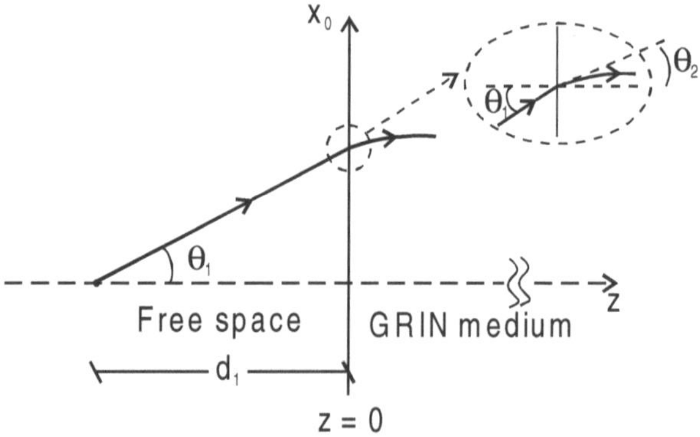

Fig. 1.5. GRIN medium illuminated by an on-axis point source at finite distance from the input plane.

In the same way, when the medium is illuminated by an on-axis point source at the input plane ($x_0 = 0$), the paraxial ray equations become

$$x(z) = H_a(z)\dot{x}_0 \tag{1.111a}$$

$$\dot{x}(z) = \dot{H}_a(z)\dot{x}_0 \tag{1.111b}$$

position x(z) and slope $\dot{x}(z)$ are now proportional to position and slope of the axial ray, respectively. These conclusions agree with definitions of the axial and field rays.

2 Imaging and Transforming Transmission Through GRIN Media

2.1 Introduction

This chapter provides a general description of light propagation through GRIN media by means of a linear integral transform and shows its use for the analysis of transformation and reconstruction of information by GRIN media. Specifically, this chapter describes paraxial optical systems with quadratic refractive index profiles through a linear integral transform called the canonical integral transform, whose kernel is expressed by the elements of the ray-transfer ABCD matrix. We also study imaging and transforming phenomena based on the properties of the canonical integral transform in ABCD systems and provide a modal representation of the kernel.

2.2 The Kernel Function

The paraxial optics system, also called the first-order or ABCD system, is a GRIN medium with quadratic transverse refractive index [2.1]. This definition includes a number of relevant discrete optical elements such as lenses and mirrors, as well as optical systems with constant refractive index. The propagation of light through first-order optical systems has been treated in Chap. 1 in the scalar and paraxial (or parabolic) approximation. This approach led to the parabolic wave equation, which has the same form as the Schrödinger equation of quantum mechanics. The first-order optical systems are described in scalar diffraction theory by a canonical integral transform. There are three ways to introduce the canonical integral transform: as an integral transform, through a matrix representation, and as a linear transformation of the Wigner function. The identity between these transforms has been established in literature [2.2–2.7]. Here the first two methods will be used because they have simple optical meaning.

Let $\psi(x_0, y_0; 0)$ be the complex amplitude of a parabolic scalar field at a transverse plane $z = 0$, then the field ψ at any plane $z > 0$ is given by the integral transform or generalized Fresnel transform [2.2–2.10]

$$\psi(x, y; z) = \int_{\Re^2} K(x_0, y_0, x, y; z)\psi(x_0, y_0; 0)\, dx_0 dy_0 \tag{2.1}$$

where K is the kernel of the linear transformation known in optics as the impulse response function or the point spread function.

The kernel is the Green's function of the parabolic wave equation (1.29) with the boundary condition [2.11]

$$K(x_0, y_0, x, y; z) \rightarrow \delta(x - x_0, y - y_0) \quad \text{as} \quad z \rightarrow 0 \tag{2.2}$$

where δ is the two-dimensional Dirac delta function.

From Eq. (2.1) it follows that the complex amplitude on an arbitrary transverse plane can be found with the aid of the kernel, if the field at a reference plane is known.

The problem now reduces to the evaluation of the kernel function for GRIN media whose dielectric constants or refractive indices are given by Eqs. (1.69). This task becomes simpler if we note that Eq. (1.38) represents the propagation of the complex amplitude of the parabolic field between two arbitrary points located at two transverse planes and joined by all classical rays, that is, by rays that minimize the optical path length. Hence Eq. (1.38) becomes the kernel satisfying Eq. (1.29) with the requirement (2.2). The connection between the kernel function and the eikonal function leads to useful relations between the geometrical ray optics and the wave optics. With this in mind, the kernel function can be written as

$$K(x_0, y_0, x, y; z) = K_0 \exp\left\{-\frac{1}{2n_0}\int_0^z \nabla_\perp^2 S^P dz'\right\} \exp\{ikS^P\} \tag{2.3}$$

where the constant K_0 is determined by the boundary condition (2.2).

The eikonal function for parabolic approximation can be expressed as

$$S^P = \int_0^z L^P dz' \tag{2.4}$$

where Eq. (1.52b) has been used, and L^P is the parabolic optical Lagrangian.

For the parabolic approximation (or in the paraxial region), the ray slopes are

$$\dot{x} \ll 1 \;, \quad \dot{y} \ll 1 \tag{2.5}$$

Expanding the optical Lagrangian, Eq. (1.52a), in a series yields

$$L^P = n_0\left[1 - \frac{g^2(z)}{2}\left(x^2 + y^2\right) + \frac{\dot{x}^2 + \dot{y}^2}{2}\right] \tag{2.6}$$

where all higher-order terms are neglected, and Eq. (1.69b) has been used.

Inserting Eq. (2.6) into Eq. (2.4) we obtain

$$S^P = n_0 z + \frac{n_0}{2}\int_0^z [\dot{x}^2 + \dot{y}^2 - g^2(z')(x^2+y^2)]dz' \tag{2.7}$$

If Eq. (1.70b) is taken into account, the integral on the right side gives

$$\int_0^z [\dot{x}^2 + \dot{y}^2 + \ddot{x}x + \ddot{y}y]dz' = \int_0^z \frac{d}{dz'}[x\dot{x} + y\dot{y}]dz' \tag{2.8}$$

Thus Eq. (2.7) reduces to

$$S^P = n_0 z + \frac{n_0}{2}(x\dot{x} + y\dot{y})|_0^z \tag{2.9}$$

Equation (2.9) can be rewritten as

$$\begin{aligned} S^P &= n_0 z + (x_0^2 + y_0^2)H_f(z)\dot{H}_f(z) + (\dot{x}_0^2 + \dot{y}_0^2)H_a(z)\dot{H}_a(z) \\ &+ 2(x_0\dot{x}_0 + y_0\dot{y}_0)H_a(z)\dot{H}_f(z) \end{aligned} \tag{2.10}$$

where Eqs. (1.77) and (1.79b) have been used, and (x_0, y_0), $(\dot{x}_0, \dot{y}_0)$ are the position and slope of any paraxial ray at z = 0.

Finally, S^P in terms of the ray position at the two reference planes is given by

$$S^P = n_0 z + \frac{n_0}{2H_a(z)}\left[(x^2+y^2)\dot{H}_a(z) + (x_0^2+y_0^2)H_f(z) - 2(x_0 x + y y_0)\right] \tag{2.11}$$

since from Eq. (1.77a) it follows that the ray slope at z = 0 can be expressed in terms of the ray position at both planes as

$$\begin{pmatrix}\dot{x}_0\\ \dot{y}_0\end{pmatrix} = \frac{\begin{pmatrix}x\\ y\end{pmatrix} - \begin{pmatrix}x_0\\ y_0\end{pmatrix}H_f(z)}{H_a(z)} \tag{2.12}$$

On the other hand, substituting Eq. (2.11) into the amplitude of the kernel function and integrating we have

$$\exp\left\{-\frac{1}{2n_0}\int_0^z \nabla_\perp^2 S dz'\right\} = \frac{1}{H_a(z)} \tag{2.13}$$

Therefore, the kernel function becomes

$$\begin{aligned} K(x_0, y_0, x, y; z) &= \frac{K_0}{H_a(z)}\exp\{ikn_0 z\}\exp\left\{i\frac{kn_0}{2H_a(z)}\right. \\ &\left.\cdot\left[(x^2+y^2)\dot{H}_a(z) + (x_0^2+y_0^2)H_f(z) - 2(x_0 x + y_0 y)\right]\right\} \end{aligned} \tag{2.14}$$

Evaluation of K_0 is carried out from condition (2.2). By using the property

$$\int_{\Re^2} \delta(x - x_0, y - y_0)\, dxdy = 1 \tag{2.15}$$

where

$$\delta(x - x_0, y - y_0) = \lim_{z\to 0} K(x_0, y_0, x, y; z) \tag{2.16}$$

we have

$$\begin{aligned} &\lim_{z\to 0} \frac{K_0}{H_a(z)} \exp\{ikn_0 z\} \exp\left\{i \frac{kn_0 H_f(z)}{2H_a(z)} (x_0^2 + y_0^2)\right\} \\ &\int_{\Re^2} \exp\left\{i \frac{kn_0}{2H_a(z)} [(x^2 + y^2)\dot{H}_a(z) - 2(xx_0 + yy_0)]\right\} dxdy = 1 \end{aligned} \tag{2.17}$$

Integration provides

$$\frac{i2\pi K_0}{kn_0} \lim_{z\to 0} \frac{1}{\dot{H}_a(z)} \exp\{ikn_0 z\} \exp\left\{i \frac{kn_0 \dot{H}_f(z)}{2\dot{H}_a(z)} (x^2 + y^2)\right\} = 1 \tag{2.18}$$

when Eqs. (1.72) and (1.74) are taken into account, K_0 is found

$$K_0 = \frac{kn_0}{i2\pi} \tag{2.19}$$

Thus the kernel function is given by

$$\begin{aligned} K(x_0, y_0, x, y; z) = &\frac{kn_0}{i2\pi H_a(z)} \exp\{ikn_0 z\} \exp\left\{i \frac{kn_0}{2H_a(z)}\right. \\ &\left. \cdot \left[\left(x^2 + y^2\right)\dot{H}_a(z) + \left(x_0^2 + y_0^2\right) H_f(z) - 2(x_0 x + y_0 y)\right]\right\} \end{aligned} \tag{2.20}$$

Equation (2.20) can also be written as

$$\begin{aligned} K(x_0, y_0, x, y; z) = &\frac{kn_0}{i2\pi H_a(z)} \exp\{ikn_0 z\} \exp\left\{i \frac{kn_0 \dot{H}_f(z)}{2H_f(z)} \left(x^2 + y^2\right)\right\} \\ &\exp\left\{i \frac{kn_0 H_f(z)}{2H_a(z)} \left[\left(x_0 - \frac{x}{H_f(z)}\right)^2 + \left(y_0 - \frac{y}{H_f(z)}\right)^2\right]\right\} \end{aligned} \tag{2.21}$$

Equations (2.20–2.21) are expressed by the elements of the ray-transfer matrix relating positions and slopes of a ray at two transverse planes as

$$K(x_0, y_0, x, y; z) = \frac{kn_0}{i2\pi B}\exp\{ikn_0 z\}$$
$$\exp\left\{i\frac{kn_0}{2B}[(x^2+y^2)D + (x_0^2+y_0^2)A - 2(x_0x + y_0y)]\right\} \tag{2.22a}$$

or

$$K(x_0, y_0, x, y; z) = \frac{kn_0}{i2\pi B}\exp\{ikn_0 z\}$$
$$\exp\left\{i\frac{kn_0 C}{2A}\left(x^2+y^2\right)\right\} \tag{2.22b}$$
$$\exp\left\{\exp\left\{\frac{ikn_0 A}{2B}\left[\left(x_0 - \frac{x}{A}\right)^2 + \left(y_0 - \frac{y}{A}\right)^2\right]\right\}\right\}$$

where Eq. (1.91a) has been used.

Equations (2.20) or (2.21) should also be parameterized by the elements of the ray-transfer matrix connecting position and optical direction cosines of a ray at two transverse planes. In particular, Eq. (2.20) is given by

$$K(x_0, y_0, x, y; z) = \frac{k}{i2\pi B'}\exp\{ikn_0 z\}$$
$$\exp\left\{i\frac{k}{2B'}\left[\left(x^2+y^2\right)D' + \left(x_0^2+y_0^2\right)A' - 2\left(x_0x + y_0y\right)\right]\right\} \tag{2.23}$$

where Eq. (1.91b) has been used.

For the above equations, we have

$$AD - BC = A'D' - B'C' = 1 \tag{2.24}$$

where Eqs. (1.90) or (1.79b) have been used.

An important particular case of the canonical integral transform is the Fresnel transform, which adequately describes diffraction in free space. For free space where $n_0 = 1$, $H_a(z) = z$, $\dot{H}_a(z) = 1$, $H_f(z) = 1$, and $\dot{H}_f(z) = 0$, Eqs. (2.20) and (2.21) become the well–known point spread function of Fresnel diffraction. It corresponds to the canonical integral transform parameterized by the transfer matrix with $A = D = 1$, $C = 0$, and $B = z$.

2.3 Imaging and Fourier Transforming Through GRIN Media

We will first review two important special cases of the generalized Fresnel transform in GRIN media, namely imaging and Fourier transforming. The imaging phenomenon is the reproduction of an arbitrary complex amplitude distribution, and Fourier transforming is the spatial spectrum of an input field.

To analyze Eq. (2.1), let us now consider planes $z = z_m$, where m is an integer, such that [2.12–2.14]

$$H_a(z_m) = 0 \quad \text{or} \quad \dot{H}_f(z_m) = 0 \tag{2.25}$$

then Eqs. (2.20–2.21) yield

$$K(x_0, y_0, x, y; z_m) = \frac{1}{H_f(z_m)} \exp\{ikn_0 z_m\} \delta\left(x_0 - \frac{x}{H_f(z_m)}, y_0 - \frac{y}{H_f(z_m)} \right) \tag{2.26}$$

Note that the quadratic phase factor

$$\exp\left\{ i \frac{kn_0 \dot{H}_f(z_m)}{2H_f(z_m)} (x^2 + y^2) \right\} \tag{2.27}$$

cancels for condition (2.25).

Hence, the complex amplitude distribution reduces to

$$\psi(x, y; z_m) = \frac{\exp\{ikn_0 z_m\}}{H_f(z_m)} \psi\left(\frac{x}{H_f(z_m)}, \frac{y}{H_f(z_m)}; 0 \right) \tag{2.28}$$

It follows from (2.28) that at planes $z = z_m$ such that $H_a(z_m) = 0$, called the imaging condition, or equivalently

$$\int_0^{z_m} g(z')\, dz' = m\pi \tag{2.29}$$

where Eq. (1.92a) has been used, the complex amplitude distribution is a replica (except for a constant factor), to the field at plane $z = 0$ with a scaling factor. In other words, the output irradiance distribution will be equal to the scaled replica of the input one.

The canonical integral transform representing such an imaging for an arbitrary input complex amplitude distribution is described by the kernel

$$K(x_0, y_0, x, y; z_m) = \frac{\exp\{ikn_0 z_m\}}{A} \delta\left(x_0 - \frac{x}{A}, y_0 - \frac{y}{A} \right) \tag{2.30}$$

where A is the magnification given by

$$A = (-1)^m \left[g_0 / g(z_m) \right]^{1/2} \tag{2.31}$$

The matrix associated with the imaging kernel is

$$M^I = \begin{pmatrix} (-1)^m \left[g_0 / g(z_m) \right]^{1/2} & 0 \\ 0 & (-1)^m \left[g(z_m) / g_0 \right]^{1/2} \end{pmatrix} \tag{2.32}$$

For GRIN media with only transverse variation of the refractive index, called selfoc media, $A = (-1)^m$ and one has nonscaled imaging of the input irradiance distribution. This corresponds to exact imaging.

On the other hand, from the imaging condition it follows that at planes $z = z_m$, the axial ray cuts the axis and the field ray is parallel to the axis. Likewise, from Eqs. (2.32) and (1.89b) it follows that the ray position at planes $z = z_m$ remains constant and is proportional to the ray position at z = 0, even if we change the slope of rays coming out from the input. The image is mapped point–to–point (Fig. 2.1).

In the same way, let us consider planes $z = \tilde{z}_p$ where p is also an integer, such that [2.12–2.14]

$$H_f(\tilde{z}_p) = 0 \quad \text{or} \quad \dot{H}_a(\tilde{z}_p) = 0 \tag{2.33}$$

Then Eq. (2.20) reduces to

$$K(x_0, y_0, x, y; \tilde{z}_p) = \frac{kn_0}{i2\pi H_a(\tilde{z}_p)} \exp\{ikn_0 \tilde{z}_p\} \exp\left\{-i \frac{kn_0}{H_a(\tilde{z}_p)} (xx_0 + yy_0)\right\} \tag{2.34}$$

According to Eqs. (2.1) and (2.34), the complex amplitude distribution at these planes is given by

$$\psi(x, y; \tilde{z}_p) = \frac{kn_0}{i2\pi H_a(\tilde{z}_p)} \exp\{ikn_0 \tilde{z}_p\} FT(f_x, f_y) \tag{2.35}$$

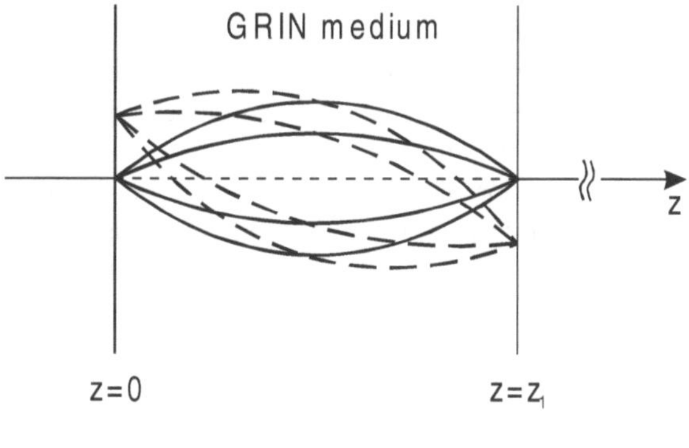

Fig. 2.1. Imaging.

where

$$\mathrm{FT}(f_x, f_y) = \int_{\Re^2} \psi(x_0, y_0; 0) \exp\{-i2\pi(f_x x_0 + f_y y_0)\} dx_0 dy_0 \tag{2.36}$$

is the ordinary Fourier transform of the input complex amplitude distribution evaluated at the spatial frequencies

$$f_x = \frac{n_0}{\lambda H_a(\tilde{z}_p)} x_0 ; \quad f_y = \frac{n_0}{\lambda H_a(\tilde{z}_p)} y_0 \tag{2.37}$$

It follows from (2.35) that at planes $z = \tilde{z}_p$ such that $H_f(\tilde{z}_p) = 0$, called the Fourier transforming condition, or equivalently

$$\int_0^{\tilde{z}_p} g(z')\, dz' = (2p+1)\frac{\pi}{2} \tag{2.38}$$

where Eq. (1.92b) has been used, the complex amplitude distribution is proportional to the spatial spectrum of the input field with a scaling factor. This means that the output irradiance is the spatial power spectrum of the input signal.

The Fourier transforming kernel is also written as

$$K(x_0, y_0, x, y; \tilde{z}_p) = \frac{kn_0}{i2\pi B} \exp\{ikn_0 \tilde{z}_p\} \exp\left\{-i\frac{kn_0}{B}(xx_0 + yy_0)\right\} \tag{2.39}$$

where B is a scaling factor given by

$$B = (-1)^p [g_0 g(\tilde{z}_p)]^{-1/2} \tag{2.40}$$

and the matrix associated with this kernel is

$$M^{FT} = \begin{pmatrix} 0 & (-1)^p [g_0 g(\tilde{z}_p)]^{-1/2} \\ (-1)^{p+1} [g_0 g(\tilde{z}_p)]^{1/2} & 0 \end{pmatrix} \tag{2.41}$$

Likewise, from the Fourier transforming condition it follows that at planes $z = \tilde{z}_p$, the field ray cuts the axis and the axial ray is parallel to the axis. If Eqs. (2.41) and (1.89b) are taken into account, we conclude that the ray slope at $z = \tilde{z}_p$ remains constant and is proportional to the ray position at z = 0. In other words, each point source of the input signal produces a parallel beam of rays at $z = \tilde{z}_p$, and on these planes we have a superposition of parallel beams with different propagation directions (Fig. 2.2). This can be regarded as the geometrical optics interpretation of the Fourier transforming phenomenon.

Finally, note that the oscillatory behavior of the elements of the ray-transfer matrix produces cyclic imaging and Fourier transforming phenomena. Consecutive inverted and erected images and power spectra are obtained for m and p odd or even integers, respectively.

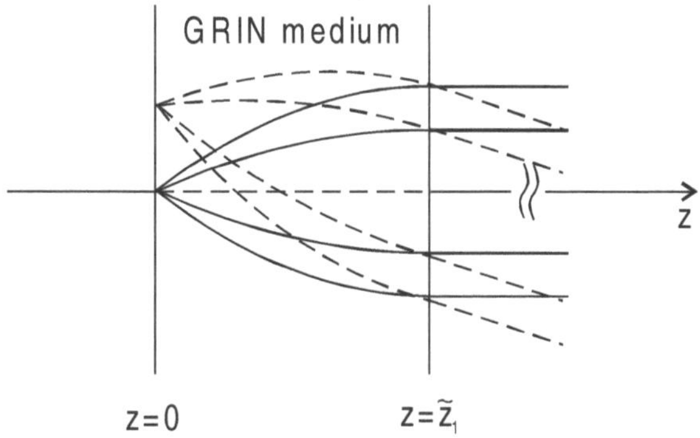

Fig. 2.2. Fourier transforming.

2.4 Fractional Fourier Transforming in GRIN Media

GRIN media play a central role in the definition of the fractional Fourier transform (FrFT) introduced into the mathematics literature by Namias [2.15], later developed by McBride and Kerr [2.16], and then applied to optics [2.7, 2.17–2.20]. The FrFT is an extension of the ordinary FT and depends on a parameter α that can be interpreted as a rotation by an angle in the space–frequency plane. The kernel of the FrFT is the propagator (Green's function) of the parabolic wave equation (1.29) in a GRIN medium that has the form of a non-stationary Schrödinger equation for the harmonic oscillator. Thus, the FrFT describes the evolution of the complex transverse amplitude during propagation through such a medium. The kernel, given by Eqs. (2.20) or (2.22), can be rewritten as

$$K_\alpha = \frac{kn_0 g_0}{i2\pi M\sin\alpha}\exp\{ikn_0 z\} \exp\left\{i\frac{kn_0 g_0}{2}\left[\frac{(x^2+y^2)}{M^2}\cot\alpha - \frac{2(xx_0+yy_0)}{M}\csc\alpha + (x_0^2+y_0^2)\cot\alpha\right]\right\} \tag{2.42}$$

where M is a scaling factor associated with the transformation and

$$\alpha = \beta \pi/2 \tag{2.43}$$

where β is the fractional order or degree of fractionality.

Likewise, Eq. (2.1) is expressed by the FrFT as

$$\mathrm{FrFT}_\alpha[\psi(x_0,y_0;0)] = \psi(x,y;z) = \int_{-\infty}^{\infty} K_\alpha \psi(x_0,y_0;0)\,dx_0 dy_0 \tag{2.44}$$

It is easy to see that for $\beta = 2m$, where m is an integer, the FrFT becomes imaging transforming and that for $\beta = 2p+1$, where p is also an integer, it reduces to ordinary Fourier transforming. Moreover, the FrFT is a periodic transform, because $\mathrm{FrFT}_{\alpha+2m\pi} = \mathrm{FrFT}_\alpha$.

The matrix associated with the fractional Fourier transform kernel K_α is

$$\begin{pmatrix} A & B \\ C & D \end{pmatrix} = \begin{pmatrix} M\cos\alpha & M\sin\alpha / g_0 \\ -g_0 \sin\alpha / M & \cos\alpha / M \end{pmatrix} \tag{2.45}$$

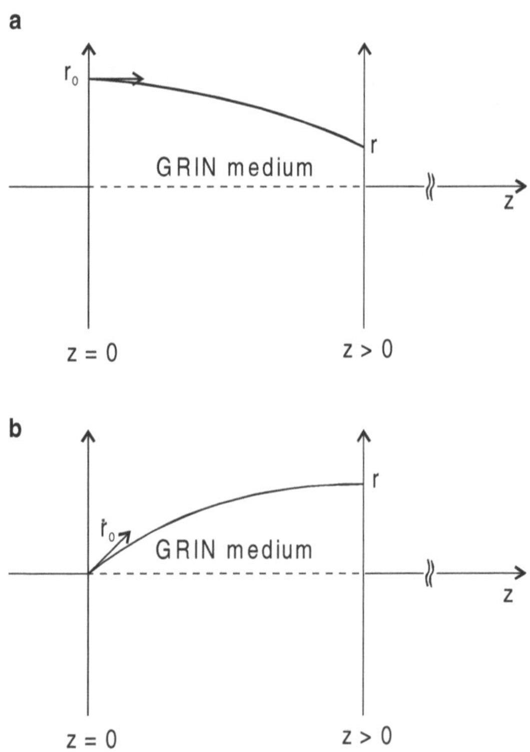

Fig. 2.3. Fractional Fourier transforming **a** $\dot{r}_0 = 0$; **b** $r_0 = 0$.

where

$$M = \sqrt{A^2 + (g_0 B)^2} = \sqrt{\frac{g_0}{g(z)}} \tag{2.46a}$$

$$\alpha = \tan^{-1}\left(\frac{g_0 B}{A}\right) = \int_0^z g(z')\,dz' \tag{2.46b}$$

and Eqs. (1.91a), (1.92), and (1.95) have been used. The scaling factor M depends on z, and α increases with z.

We will now examine the interpretation of the FrFT from a ray optics viewpoint. Let two particular paraxial rays be characterized by their position and slope at the input plane, taking into account that any paraxial ray in a GRIN medium can be regarded as a linear combination of two rays, one parallel to the optical axis at the input plane ($\dot{r}_0 = 0$), the other emanating from the axial point at the input plane ($r_0 = 0$). The position of these two rays at any z > 0 can, respectively, be expressed by

$$r(z) = r_0 A = r_0 M \cos\alpha \tag{2.47a}$$

$$r(z) = \dot{r}_0 B = \dot{r}_0 \frac{M\sin\alpha}{g_0} \tag{2.47b}$$

where Eqs. (1.91a), (1.106a), (1.107a), and (2.45) have been used.

From Eqs. (2.47) it follows that the ratio of the position of the output ray to the position of the input ray is

$$\frac{r(z)}{r_0} = A = M\cos\alpha \tag{2.48a}$$

as shown in Fig. 2.3a, and the ratio of the position of the output ray to the slope of the input ray is

$$\frac{r(z)}{\dot{r}_0} = B = \frac{M\sin\alpha}{g_0} \tag{2.48b}$$

as shown in Fig. 2.3b. Then, M and α (or β) given by (2.46) can be determined by using (2.48).

Equations (2.48) are the key to interpreting light propagation in the space–frequency plane, as shown in Fig. 2.4. The horizontal axis r/r_0 is the imaging axis obtained as $B = H_a = 0$ (imaging condition), and the vertical axis $g_0 r/\dot{r}_0$ is the ordinary Fourier transforming axis obtained as $A = H_f = 0$ (transforming condition). In this plane, M and α denote polar coordinates for propagation through a GRIN medium. At any z > 0, M$\cos\alpha$ and M$\sin\alpha$ correspond to the projection of M to the imaging and the ordinary Fourier transforming axes, respectively (Fig. 2.4). Then α, which increases with z, represents an angle in the space–frequency plane. For $\alpha = m\pi$, we have a set of erected and inverted images with magnification $\sqrt{g_0/g(z_m)}$, and for $\alpha = (2p+1)\pi/2$ a set of erected and inverted ordinary Fourier spectra with magnification

$$\frac{1}{g_0}\sqrt{\frac{g_0}{g(\tilde{z}_p)}} = [g_0 g(\tilde{z}_p)]^{-1/2}$$

is obtained.

In the simple example of a z–independent index gradient (selfoc medium), one has $\alpha = g_0 z$, i.e., the angle α is proportional to the propagation distance z. In a selfoc medium, the FT with fractional order β is obtained at $z = \beta L$, where $L = \pi/2g_0$ is the GRIN length for the ordinary Fourier transform.

Likewise, the FrFT concept has been extended to the anamorphic GRIN case by using an elliptic GRIN medium whose dielectric constant is given by

$$n^2(x,y,z) = n_0^2[1 - g_x^2(z)x^2 - g_y^2(z)y^2] \tag{2.49}$$

where g_x, g_y are the gradient parameters along x–, y– axis.

This modification permits the use of different fractional orders for two orthogonal axes of a two–dimensional input field.The anamorphic case has also been widely studied by conventional cylindrical lenses for applications in signal processing. The main applications are related to changes of space variance, chirp–noise removal, and signal multiplexing [2.21–2.23].

Finally, we calculate the maximum length of a GRIN medium permissible for obtaining imaging and transforming transmission without an estimable loss of information because of aberrations.

The refractive index of a more general rotationally symmetric GRIN medium up to fourth order can be expressed as

$$n(x,y,z) = n_0\left[1 - \frac{g^2(z)}{2}(x^2+y^2) - \frac{g^4(z)}{8}(x^2+y^2)^2\right] \tag{2.50}$$

Under such a condition, the paraxial optical path given by Eq. (2.7) becomes

$$S \approx S^P - \frac{n_0}{8}\int_0^z g^4(z')(x^2+y^2)^2\,dz' \tag{2.51}$$

for primary aberrations.

It is evident that the fourth and higher–order terms do not appreciably contribute to the aberrations if

$$k|S - S^P| \ll \pi \tag{2.52}$$

Substituting Eq. (2.51) into condition (2.52), we have

$$\left|\int_0^{z_{max}} g^4(z')\,dz'\right| \ll \frac{4\lambda_0}{n_0 r_0^4} \tag{2.53}$$

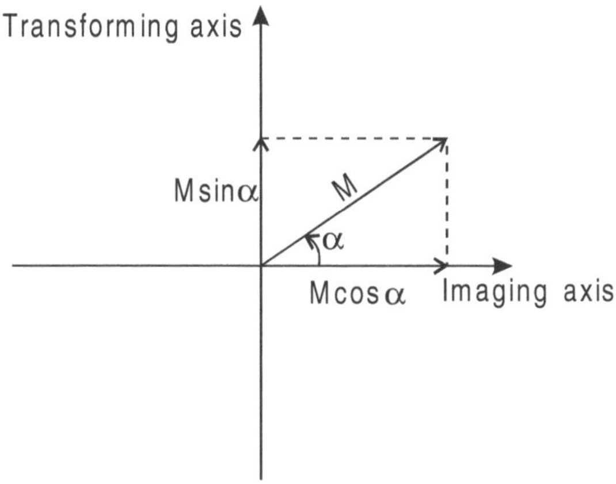

Fig. 2.4. Space–frequency plane for propagation through a GRIN medium.

where $r_0 = \sqrt{x_0^2 + y_0^2}$ is the semiaperture of the GRIN medium, and z_{max} is the maximum length.

Equation (2.53) may be used for calculating the maximum length of a GRIN medium for perfect image and transform transmission. As an example, the maximum length will be evaluated for a selfoc medium with a constant gradient parameter g_0. For this selfoc medium, from Eq. (2.53) it follows that

$$z_{max} \ll \frac{4\lambda_0}{n_0 (g_0 r_0)^4} \tag{2.54}$$

Then for typical values $\lambda_0 = 0.75\,\mu m$, $n_0 = 1.5$, $g_0 = 0.1\,mm^{-1}$, and $r_0 = 0.2\,mm$, we obtain that $z_{max} \le 1.25\,m$ if a factor of 10 is considered a good criterion for "much less than."

2.5 Modal Representation of the Kernel

We will consider the problem of modal propagation of a monochromatic scalar field through a GRIN medium whose dielectric constant is given by Eq. (1.69a), that is

$$n^2(x, y, z) = n_0^2[1 - g^2(z)(x^2 + y^2)] \tag{2.55}$$

For waves propagating in a direction close to the z-axis, the complex amplitude of the scalar field can be written as

$$\psi(\vec{r}) = \phi(\vec{r}) e^{ikn_0 z} \tag{2.56}$$

where the variation of ϕ with position is slow within the distance of a wavelength in the z-direction.

For this case, the parabolic wave equation is expressed as

$$\nabla_{\perp}^{2}\phi + 2ikn_{0}\frac{\partial\phi}{\partial z} + k^{2}(n^{2} - n_{0}^{2})\phi = 0 \qquad (2.57)$$

It can be rewritten as

$$2ikn_{0}\frac{\partial\phi}{\partial z} = H\phi \qquad (2.58)$$

where H is the Hamilton's operator given by

$$H = -\nabla_{\perp}^{2} - k^{2}(n^{2} - n_{0}^{2}) = -\nabla_{\perp}^{2} + k^{2}n_{0}^{2}g^{2}(z)(x^{2} + y^{2}) \qquad (2.59)$$

The Hamiltonian operator satisfies the eigenvalue equation

$$H\phi_{h,j} = E_{h,j}\phi_{h,j} \qquad (2.60)$$

where $\phi_{h,j}$ is an eigenfunction of H, and $E_{h,j}$ is the corresponding eigenvalue with h, j natural numbers.

Equation (2.60) with (2.59) has the same form as the Schrödinger equation for the two–dimensional harmonic oscillator; its eigenmodes at any $z > 0$ are [2.24]

$$\psi_{h,j}(x, y; z) = \phi_{h,j}(x, y; z)e^{i\beta_{h,j}z} = \phi_{h}(x; z)\phi_{j}(y; z)e^{i\beta_{h,j}z} \qquad (2.61)$$

where

$$\phi_{\sigma}(t; z) = \left[\frac{\mu(z)}{\pi^{1/2}2^{\sigma}\sigma!}\right]^{1/2} \exp\left\{-\frac{\mu^{2}(z)t^{2}}{2}\right\}H_{\sigma}[\mu(z)t] \qquad (2.62a)$$

with $\sigma = h, j$; $t = x, y$; H_{σ} the Hermite polynomial of order σ;

$$\mu(z) = [kn_{0}g(z)]^{1/2} \qquad (2.62b)$$

the inverse of the width or of the spot size of the Gaussian function; and

$$\beta_{h,j} = kn_{0} - \frac{(h + j + 1)}{z}\int_{0}^{z} g(z')\,dz' \qquad (2.62c)$$

the propagation constant of the h, jth mode.

The Hermite polynomials are defined by

$$H_{\sigma}[\mu(z)t] = (-1)^{\sigma}\exp\{\mu^{2}(z)t^{2}\}\frac{1}{\mu^{\sigma}}\frac{d^{\sigma}}{dt^{\sigma}}\exp\{-\mu^{2}(z)t^{2}\} \qquad (2.63)$$

For the lowest orders we have

$$H_{0}(\mu t) = 1;\ H_{1}(\mu t) = 2\mu t;\ H_{2}(\mu t) = 4\mu^{2}t^{2} - 2;\ H_{3}(\mu t) = 8\mu^{3}t^{3} - 12\mu t \qquad (2.64)$$

In the same way, eigenvalues of Eq. (2.60) are given by

$$E_{h,j} = 2\mu^2(z)(h + j + 1) \tag{2.65}$$

where recursion formulas for Hermite polynomials

$$\frac{dH_\sigma(\mu t)}{dt} = 2\sigma\mu H_{\sigma-1}(\mu t) \tag{2.66a}$$

and

$$H_{\sigma+1}(\mu t) = 2\mu t H_\sigma(\mu t) - 2\sigma H_{\sigma-1}(\mu t) \tag{2.66b}$$

have been used.

From Eqs. (2.60–2.62) it follows that the eigenmodes are the well–known normalized Hermite–Gaussian functions, and thus ψ may be expressed in terms of these eigenmodes as

$$\psi(x, y; z) = \sum_{h,j} c_{h,j}(z)\psi_{h,j}(x, y; z) \tag{2.67}$$

since they constitute a complete orthonormal and discrete set, that is,

$$\int_{\Re^2} \psi_{h,j}(x, y; z)\psi^*{}_{h',j'}(x, y; z)\, dxdy = \delta_{hjh'j'} \tag{2.68}$$

where $\delta_{hjh'j'}$ is the Kronecker's delta symbol that equals unity if $h = h'$, $j = j'$, and is zero otherwise.

The coefficients in the modal expansion (2.67) can be determined by multiplying both sides by $\psi^*_{h,j}$, integrating, and taking into account Eq. (2.68), that is

$$c_{h,j}(z) = \int_{\Re^2} \psi(x, y; z)\psi^*_{h,j}(x, y; z)\, dxdy \tag{2.69}$$

The modal expansion can be conveniently performed by using Green's function K, which also satisfies Eq. (2.58) with the initial condition

$$K(x_0, y_0, x, y; 0) = \delta(x - x_0, y - y_0)$$

With the help of this Green's function, we can directly link the scalar field at two different planes by the integral equation (2.1). Using the orthonormality properties of the eigenmodes of the Hamiltonian H, one can show that K is given by

$$K(x_0, y_0, x, y; 0) = \sum_{h,j} \psi_{h,j}(x, y; z)\psi^*_{h,j}(x_0, y_0; 0) \tag{2.70}$$

Substitution of Eqs. (2.61–2.62a) and (2.62c) into Eq. (2.70) provides

$$K = \frac{\mu\mu_0}{\pi}\exp\left\{i\left[kn_0 z - \int_0^z g(z')dz'\right]\right\}\exp\left\{-\frac{1}{2}\left[\mu^2\left(x^2+y^2\right)+\mu_0^2\left(x_0^2+y_0^2\right)\right]\right\}\cdot$$
$$\sum_h \frac{H_h(\mu x)H_h(\mu_0 x_0)}{2^h h!}\exp\left\{-ih\int_0^z g(z')dz'\right\}\cdot \tag{2.71}$$
$$\sum_j \frac{H_j(\mu y)H_j(\mu_0 y_0)}{2^j j!}\exp\left\{-ij\int_0^z g(z')dz'\right\}$$

where μ_0 is the value of $\mu(z)$ at $z = 0$, and x_0, y_0 are coordinates at the input plane.

Clearly, if we make the substitution of Eq. (2.46b)

$$\alpha = \int_0^z g(z')\,dz' \tag{2.72}$$

in Eq. (2.71), we obtain the modal representation of the transforming kernel

$$K_\alpha = \frac{\mu\mu_0}{\pi}\exp\{i[kn_0 z - \alpha]\}\exp\left\{-\frac{1}{2}\left[\mu^2\left(x^2+y^2\right)+\mu_0^2\left(x_0^2+y_0^2\right)\right]\right\}$$
$$\cdot\sum_h \frac{H_h(\mu x)H_h(\mu_0 x_0)}{2^h h!}\exp\{-ih\alpha\}\sum_j \frac{H_j(\mu y)H_j(\mu_0 y_0)}{2^j j!}\exp\{-ij\alpha\} \tag{2.73}$$

which has considerable application to the theory of laser resonator modes [2.25].

From Eq. (2.73) it follows that for $\alpha = m\pi$ (imaging condition), $K_{m\pi}$ becomes the delta Dirac function, and at z_m planes all the modes h,j are in the same phase relationship as in the input plane. Likewise, for $\alpha = (2p+1)\pi/2$ (transforming condition), $K_{(2p+1)\pi/2}$ reduces to Eq. (2.34), and at $\tilde{z}_p$ planes all modes h,j are in quadrature with the input ones.

However, in GRIN media it is more practical to obtain a parameterized representation of the kernel by the elements of the ray-transfer matrix, which facilitates evaluation of the integral representation for fractional transforms. By using Mehler's formula for the Hermite polynomials

$$\sum_\sigma \frac{\exp\{-i\sigma\alpha\}\exp\left\{-\frac{1}{2}\left(\mu_0^2 t_0^2+\mu^2 t^2\right)\right\}}{2^\sigma \sigma!}H_\sigma(\mu t)H_\sigma(\mu_0 t_0) = [1-\exp\{-i2\alpha\}]^{-1/2}$$
$$\exp\left(\frac{4\mu\mu_0 t t_0\exp\{-i\alpha\} - \left[\mu^2 t^2+\mu_0^2 t_0^2\right][1+\exp\{-i2\alpha\}]}{2[1-\exp\{-i2\alpha\}]}\right) \tag{2.74}$$

modal representation of the kernel becomes

$$K_\alpha = \frac{\mu\mu_0 \exp\{i(kn_0 z - \alpha)\}}{\pi[1-\exp\{-i2\alpha\}]}\exp\left\{\frac{1}{1-\exp\{-i2\alpha\}}\left[2\mu\mu_0(xx_0+yy_0)\exp\{-i\alpha\} - \frac{[\mu_0^2(x_0^2+y_0^2)+\mu^2(x^2+y^2)][1+\exp\{-i2\alpha\}]}{2}\right]\right\} \tag{2.75}$$

We may now write

$$1-\exp\{-i2\alpha\} = i2\sin\alpha\exp\{-i\alpha\} \tag{2.76a}$$

$$\frac{2\exp\{-i\alpha\}}{1-\exp\{-i2\alpha\}} = -i\csc\alpha \tag{2.76b}$$

$$\frac{1+\exp\{-i2\alpha\}}{1-\exp\{-i2\alpha\}} = -i\cot\alpha \tag{2.76c}$$

Inserting Eqs. (2.76) into Eq. (2.75) we obtain

$$K_\alpha(x_0,y_0,x,y;z) = \frac{kn_0 g_0}{i2\pi M\sin\alpha}\exp\{ikn_0 z\}\exp\left\{i\frac{kn_0 g_0}{2}\left[\frac{(x^2+y^2)}{M^2}\cot\alpha - \frac{2(xx_0+yy_0)}{M}\csc\alpha + (x_0^2+y_0^2)\cot\alpha\right]\right\} \tag{2.77}$$

where Eqs. (2.46a) and (2.63) have been used. This result based on the modal analysis, coincides with the result obtained in Sect. 2.3.

Finally, as mentioned in Sect. 2.4, in the paraxial approximation perfect image and transform transmission result through GRIN media. However, one should remember that this result is valid exclusively within the paraxial approximation. Direct solution of the scalar wave equation without resort to the parabolic approximation provides a different conclusion, and we can only determine the maximum distance because it is necessary to take modal dispersion into account. In this case, the only difference is in the propagation constant of the h,j mode, which is now given by

$$\beta'_{hj} = kn_0\left[1 - \frac{2(h+j+1)}{kn_0 z}\int_0^z g(z')\,dz'\right]^{1/2} = kn_0\left[1 - \frac{2(h+j+1)}{kn_0 z}\alpha\right]^{1/2} \tag{2.78}$$

The propagation constant is not proportional to α as in Eq. (2.64). It thus follows that, for instance, there no longer exist planes in which all the modes h,j are in the same phase relationship as in the input plane. In other words, it will be necessary to take into account the modal dispersion through a GRIN material in order to calculate maximum distance for which information can be transmitted without appreciable distortion.

Expanding Eq. (2.78) in a power series, we have

$$\beta'_{hj} = \beta_{hj} - \frac{(h+j+1)^2 \alpha^2}{2kn_0 z^2} + \ldots \tag{2.79}$$

It is evident that the second and higher–order terms do not significantly contribute to the modal dispersion if

$$|\beta'_{hj} - \beta_{hj}| z \langle\langle \pi \tag{2.80}$$

Then the maximum length z_m of a GRIN medium to perfect image and transform transmission can be expressed as

$$z_{max} \rangle\rangle \frac{\lambda_0 (h+j+1)^2_{max} \alpha^2_{max}}{4\pi^2 n_0} \tag{2.81}$$

Let us give a numerical example for a selfoc medium. In this medium $\alpha_{max} = g_0 z_{max}$, and from Eq. (2.81) we have

$$z_{max} \langle\langle \frac{4\pi^2 n_0}{g_0^2 \lambda_0 (h+j+1)^2_{max}} \tag{2.82}$$

We obtain $z_{max} \leq 1.26\,\mathrm{m}$ for $(h+j+1)^2_{max} = 25$ and typical values $\lambda_0 = 0.75\,\mu\mathrm{m}$, $n_0 = 1.5$, and $g_0 = 0.1\,\mathrm{mm}^{-1}$.

In short, the consideration leading to Eq. (2.81) in GRIN media has the same physical origin as those used to describe group velocity dispersion and its influence on pulse broadening in waveguides in the approximation that material dispersion is neglected. In the latter case, group velocity dispersion manifests itself as a limitation on the number of pulses that can be transmitted per unit time through a maximum distance [2.26].

3 GRIN Lenses for Uniform Illumination

3.1 Introduction

A GRIN lens consists of a cylinder of inhomogeneous dielectric material with a refractive index distribution that has a maximum at the cylinder axis and decreases continuously from the axis to the periphery along the transverse direction. The focusing and transforming capabilities of GRIN lenses come from a quadratic variation in refractive index with radial distance from the axis. In a GRIN lens, rays follow sinusoidal trajectories as if they were curved (bent) by a force toward the higher refractive index. The imaging and transforming rules of GRIN lenses are, in principle, the same as those for homogeneous lenses, but there are differences in behavior between them. For instance, the focal length of a GRIN lens only depends on the lens thickness assuming planar input and output faces. This means that a GRIN lens can behave as a convergent, divergent, or telescopic lens depending on its thickness. Therefore, it is important that GRIN lenses be designed properly to perform typical functions as on–axis and off–axis imaging, collimation, and focusing in optical systems.

In this chapter transmittance function, lens law, geometrical optics, and the effect of finite size of a GRIN lens on image and Fourier transform formation will be studied and discussed when a GRIN lens is illuminated by a uniform monochromatic wave.

3.2 Transmittance Function of a GRIN Lens for Uniform Illumination

A GRIN lens can be considered as a segment of a cylindrical GRIN medium of radius a and length d limited by plane parallel faces perpendicular to the axis whose refractive index profile is given by

$$n^2(x,y,z)=\begin{cases} n_0^2[1-g^2(z)(x^2+y^2)] & \text{for } r=(x^2+y^2)^{1/2}\le a \text{ and } 0\le z\le d \\ 1 & \text{otherwise} \end{cases} \tag{3.1}$$

provided that the GRIN lens is surrounded by vacuum or free space, where n_0 is the index at the lens axis and g(z) is the gradient parameter (Fig. 3.1). Depending on the functional form of g(z), different kinds of radial and axial GRIN lenses can be found. For instance, selfoc lenses are obtained for g(z) constant [3.1]; tapered lenses [3.2] for $g(z)=g_0/(1\pm z/L)^q$, where q is a real number, $g_0=g(0)$, and L a parameter; exponential lenses [3.3] for $g(z)=g_0 e^{\pm z/L}$, and so on. GRIN lenses are usually used in miniature optical systems, micro–optics devices, optical fiber transmission systems, and optical sensors.

First we will study the transmittance function of a GRIN lens for uniform illumination, then we will determine the focal length of this lens. It is well–known that when a lens is illuminated by a wavefront, the transmittance function is defined as the ratio between the complex amplitude distributions at the end surfaces. We now suppose a GRIN lens illuminated by a uniform monochromatic wavefront such as a spherocylindrical wavefront of curvature radii d_1 and d_2. The complex amplitude distributions at the input face z = 0 due to spherocylindrical illumination in the paraxial region can be written, apart from a constant, as

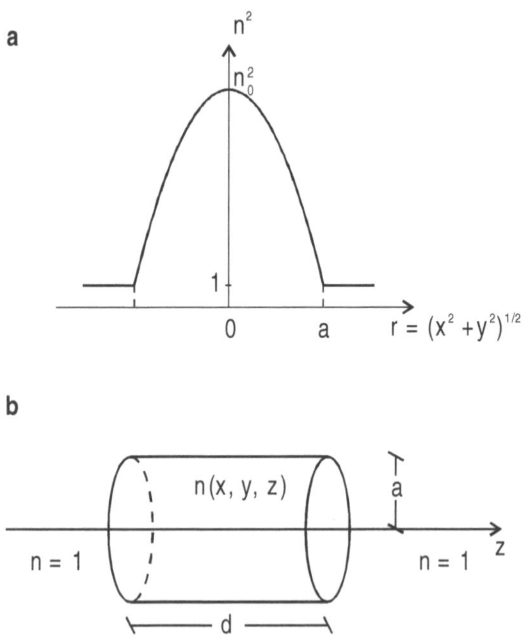

Fig. 3.1. GRIN lens: **a** transverse refractive index and **b** geometry.

$$\psi(x_0, y_0; 0) = \frac{1}{(d_1 d_2)^{1/2}} \exp\left\{ i \frac{\pi}{\lambda} \left(\frac{x_0^2}{d_1} + \frac{y_0^2}{d_2} \right) \right\} \tag{3.2}$$

and the complex amplitude distribution $\psi(x,y;d)$ at the output face $z = d$ of the GRIN lens will be given by Eq. (2.1).

Thus, the transmittance function of the GRIN lens is expressed as

$$t(x_0, y_0, x, y; d) = \frac{\psi(x, y; d)}{\psi(x_0, y_0; 0)} = \frac{\int_{\Re^2} K(x_0, y_0, x, y; d)\, \psi(x_0, y_0; 0)\, dx_0 dy_0}{\psi(x_0, y_0; 0)} \tag{3.3}$$

where Eq. (2.1) has been used, and K is the impulse response of the GRIN lens given by Eq. (2.20).

By substituting Eqs. (2.20) and (3.2) into the numerator of Eq. (3.3) and integrating, we have

$$\begin{aligned} &\frac{n_0 \exp\{ikn_0 d\}}{\{[H_a(d) + n_0 d_1 H_f(d)][H_a(d) + n_0 d_2 H_f(d)]\}^{1/2}} \\ &\cdot \exp\left\{ i \frac{\pi n_0}{\lambda} \left[\frac{\dot{H}_a(d) + n_0 d_1 \dot{H}_f(d)}{H_a(d) + n_0 d_1 H_f(d)} x^2 + \frac{\dot{H}_a(d) + n_0 d_2 \dot{H}_f(d)}{H_a(d) + n_0 d_2 H_f(d)} y^2 \right] \right\} \end{aligned} \tag{3.4}$$

When expression (3.4) and Eq. (3.2) are inserted into Eq. (3.3), the transmittance function for uniform spherocylindrical illumination can be written as

$$\begin{aligned} t(x_0, y_0, x, y; d) = n_0 &\left\{ \frac{d_1 d_2}{[H_a(d) + n_0 d_1 H_f(d)][H_a(d) + n_0 d_2 H_f(d)]} \right\}^{1/2} \\ &\cdot \exp\{ikn_0 d\} \exp\left\{ -i \frac{\pi}{\lambda} \left(\frac{x_0^2}{d_1} + \frac{y_0^2}{d_2} \right) \right\} \\ &\cdot \exp\left\{ i \frac{\pi n_0}{\lambda} \left[x^2 \frac{d}{dz} \left(\ln[H_a(z) + n_0 d_1 H_f(z)] \Big|_{z=d} \right) \right.\right. \\ &\left.\left. + y^2 \frac{d}{dz} \left(\ln[H_a(z) + n_0 d_2 H_f(z)] \Big|_{z=d} \right) \right] \right\} \end{aligned} \tag{3.5}$$

Note that the above equation for transmittance function can also be obtained from the kernel or impulse response of the lens if the kernel definition, with requirement (2.2), is slightly reformulated by the boundary condition [3.4]

$$K(x_0, y_0, x, y; d) \to 1 \quad \text{as} \quad d \to 0 \tag{3.6}$$

Likewise, for $d_1 = d_2 = d_0$ (uniform spherical wavefront), the transmittance function becomes

$$t(r_0, r; d) = \frac{n_0 d_0}{H_a(d) + n_0 d_0 H_f(d)} \exp\{ikn_0 d\} \exp\left\{-i\frac{\pi r_0^2}{\lambda d_0}\right\} \cdot \exp\left\{i\frac{\pi n_0 r^2}{\lambda}\frac{d}{dz}\left(\ln[H_a(z) + n_0 d_0 H_f(z)]\big|_{z=d}\right)\right\} \tag{3.7}$$

where $r_{(0)}^2 = x_{(0)}^2 + y_{(0)}^2$.

The transmittance function for a uniform plane wavefront is obtained from the last equation making $d_0 \to \infty$, that is

$$t(r; d) = \frac{\exp\{ikn_0 d\}}{H_f(d)} \exp\left\{i\frac{\pi n_0 r^2}{\lambda}\frac{d}{dz}(\ln H_f(z)|_{z=d})\right\} \tag{3.8}$$

In this case, to identify the nature of the transmittance function we must examine the dependence on r. Since only a quadratic term in r is presented, the transmittance function may be regarded as a parabolic approximation to a spherical wave converging toward (or diverging from) a point that lies on the z–axis. Comparing Eq. (3.8) with that of a spherical wave converging toward (or diverging) from a point at length l' to the plane $z = d$

$$\exp\left\{-i\frac{\pi}{\lambda l'}r^2\right\} \tag{3.9}$$

we find that

$$l'(d) = -\left[n_0 \frac{d}{dz}\ln H_f(z)|_{z=d}\right]^{-1} = -\frac{H_f(d)}{n_0 \dot{H}_f(d)} = \frac{1}{n_0 g^2(z)}\frac{d}{dz}\ln \dot{H}_f(z)|_{z=d} \tag{3.10}$$

where Eq. (1.70b) has been used.

When l' is positive, the wavefront is converging toward a point located to the right of the output face $z = d$, and when l' is negative, the wavefront is diverging from a point situated to the left of plane $z = d$ (Fig. 3.2).

Thus a GRIN lens (rod lens) can be considered as a conventional spherical lens with an equivalent back focal length measured from the output face, given by Eq. (3.10), also called the working distance. This focal length depends on the thickness of the GRIN lens.

The working distance can be written as

$$l'(d) = \frac{1}{n_0 g(d)} \operatorname{cotan}\left[\int_0^d g(z')\,dz'\right] \tag{3.11}$$

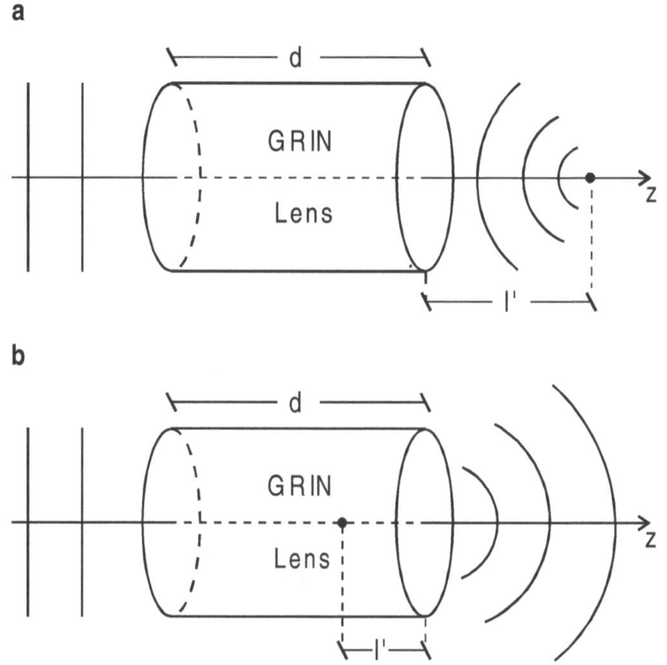

Fig. 3.2. Equivalent focal length of a GRIN lens for **a** $l'>0$, **b** $l'<0$.

where Eqs. (1.92b) and (1.95b) have been used.

We now examine GRIN lens behavior from a ray optics viewpoint. When the GRIN lens is illuminated by a collimated light beam, ray position and ray slope inside the lens are given by Eqs. (1.110). Then, for exiting paraxial ray heights and slopes we have

$$r(d) = r_0 H_f(d) \tag{3.12a}$$

$$\dot{r}(d) = n_0 r_0 \dot{H}_f(d) \tag{3.12b}$$

After propagation of light in free–space a distance d', position and slope of rays on the plane located at d' from the output face of the GRIN lens, as shown in Fig. 3.3, are given by

$$r(d+d') = r(d) + d'\dot{r}(d) = r_0\left[H_f(d) + n_0 d'\dot{H}_f(d)\right] \tag{3.13a}$$

$$\dot{r}(d+d') = \dot{r}(d) \tag{3.13b}$$

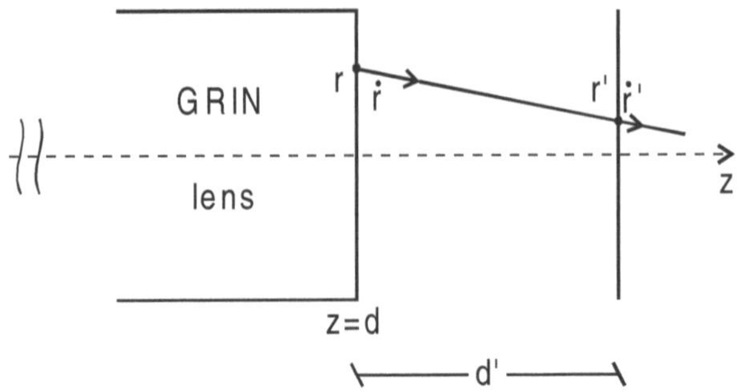

Fig. 3.3. Free–space propagation.

The ray positions at the back focus lie on the z-axis for any value of ray slopes. Thus the condition $r(d+d')=0$ independent of $\dot{r}(d+d')$ can be expressed as

$$d' = l'(d) = -\frac{H_f(d)}{n_0 \dot{H}_f(d)} \tag{3.14}$$

where d' is, in this case, the back working distance. Equation (3.14) coincides with Eq. (3.10). Note that both equations can be parameterized by the elements of the ray–transfer matrix of the lens and rewritten as

$$l' = -\frac{A}{n_0 C} = -\frac{A'}{C'} \tag{3.15}$$

where Eqs. (1.91) have been used.

For a selfoc lens, Eq. (3.11) or (3.14) reduces to

$$l'(d) = \frac{\text{cotan}(g_0 d)}{n_0 g_0} \tag{3.16}$$

and the power is given by

$$P'(d) = n_0 g_0 \tan(g_0 d) \tag{3.17}$$

If the thickness of the selfoc lens is thin enough so that $g_0 d \langle\langle 1$, i.e., a Wood lens, the power is $n_0 g_0^2 d$.

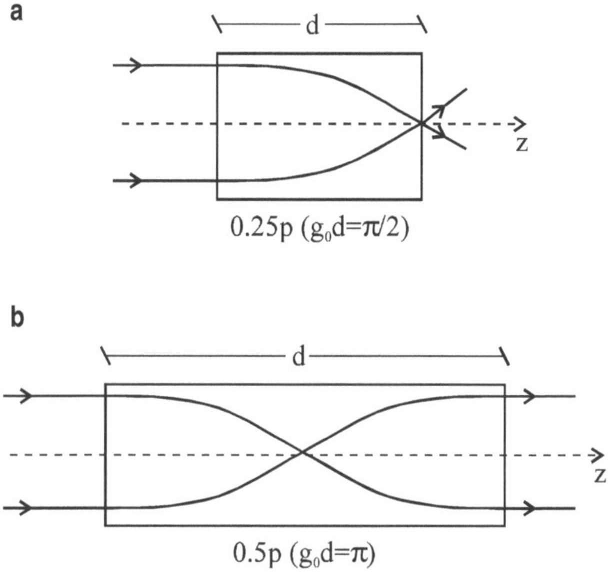

Fig. 3.4. a Quarter-pitch selfoc lens forms an inverted, real image of an object at the infinite on the output face of the lens (focusing selfoc lens), **b** Half-pitch selfoc lens transfers an inverted real image from one face of the lens to the other (telescopic selfoc lens).

In a selfoc lens illuminated by a collimated light beam, ray equations through the lens are given by

$$r(z) = r_0 \cos(g_0 z) \tag{3.18a}$$

$$\dot{r}(z) = -r_0 g_0 \sin(g_0 z) \tag{3.18b}$$

where Eqs. (1.96) and (1.110) have been used.

Light oscillates about the axis of the lens with a period $2\pi/g_0$, known as the pitch. Knowing the pitch, it is possible to explain the behavior of a selfoc lens by varying the length of the lens. For quarter-pitch $(g_0 d = \pi/2)$ and three-quarter pitch $(g_0 d = 3\pi/2)$ lenses, the focus is located on the output face, and for half-pitch $(g_0 d = \pi)$ and one-pitch $(g_0 d = 2\pi)$, the focus is at infinity. Figure 3.4 shows quarter-pitch and half-pitch selfoc lenses. Likewise, if the length of the selfoc lens is such that $0 < g_0 d < \pi/2$ and $\pi < g_0 d < 3\pi/2$, a real focus is obtained. On the contrary, if $\pi/2 < g_0 d < \pi$ and $3\pi/2 < g_0 d < 2\pi$, a virtual focus results. The working distances of 0.23– and 0.29–pitch selfoc lenses with $n_0 = 1.602$ and $g_0 = 0.192\,\text{mm}^{-1}$ are 0.41 mm and -0.83 mm, respectively.

3.3 GRIN Lens Law: Imaging and Fourier Transforming by GRIN Lens

In Sect. 3.2, it was shown that a GRIN lens with revolution symmetry may be regarded as a conventional spherical lens with an equivalent focal length. In this section, based on this result, we study image and transform formation by a GRIN lens for uniform illumination. Referring to the geometry of Fig. 3.5, a planar object is placed a distance d_1 in front of a GRIN lens of length d and is illuminated by a uniform spherical wave of wavelength λ diverging from an on–axis point source. At a distance d_1' behind the GRIN lens there is an observation plane [3.5]. Our aim is to find the conditions in which the output of the optical system can be said to be an image or a Fourier transform of the object.

Let us consider a complex amplitude transparency $f_s(x_1, y_1)$. When it is illuminated by a uniform monochromatic spherical wave of curvature radius d_0, the complex amplitude distribution on the input plane of the optical system can be written as

$$\psi_i(x_1, y_1) = \frac{1}{d_0} \exp\left\{ i \frac{\pi}{\lambda d_0} (x_1^2 + y_1^2) \right\} f_s(x_1, y_1) \tag{3.19}$$

Neglecting finite dimensions of the GRIN lens in a first approach, the complex amplitude distribution at the output plane is given by Eq. (2.1), that is

$$\psi_0(x_1', y_1') = \int_{\Re^2} K(x_1, y_1, x_1', y_1'; d_1', d, d_1)\, \psi_i(x_1, y_1)\, dx_1 dy_1 \tag{3.20}$$

where K is the kernel or the impulse response of the optical system.

Evaluation of K is carried out from the elements of the ray–transfer matrix of the optical system. The total matrix corresponding to the optical system shown in Fig. 3.5 is given by the product of the ray matrices for free spaces d_1 and d_1', and for a GRIN lens of thickness d, that is

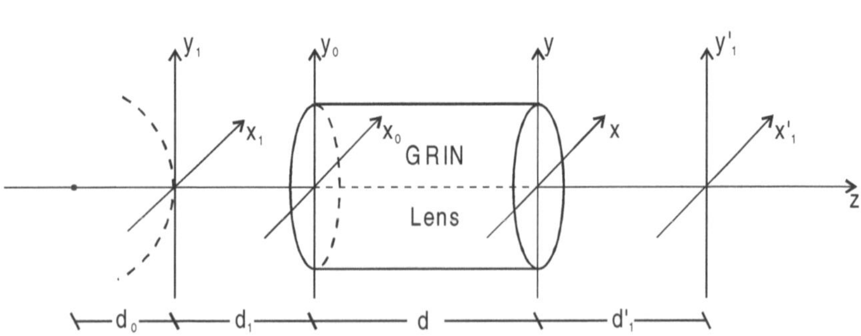

Fig. 3.5. Geometry for image and transform formation by GRIN lens.

$$M_s = \begin{pmatrix} 1 & d_1' \\ 0 & 1 \end{pmatrix} \begin{pmatrix} H_f(d) & H_a(d)/n_0 \\ n_0\dot{H}_f(d) & \dot{H}_a(d) \end{pmatrix} \begin{pmatrix} 1 & d_1 \\ 0 & 1 \end{pmatrix} =$$

$$= \begin{pmatrix} H_f(d)+n_0 d_1'\dot{H}_f(d) & \frac{1}{n_0}\{H_a(d)+n_0 d_1 H_f(d)+n_0 d_1'[\dot{H}_a(d)+n_0 d_1 \dot{H}_f(d)]\} \\ n_0\dot{H}_f(d) & \dot{H}_a(d)+n_0 d_1 \dot{H}_f(d) \end{pmatrix}$$

$$= \begin{pmatrix} A_s' & B_s' \\ C_s' & D_s' \end{pmatrix} \tag{3.21}$$

From Eq. (3.21) it follows that K can be expressed as

$$K(x_1, y_1, x_1', y_1'; d_1', d, d_1) = \frac{k}{i2\pi B_s'} \exp\left\{ i\frac{k}{2B_s'} \cdot \left[(x_1'^2 + y_1'^2) D_s' + (x_1^2 + y_1^2) A_s' - 2(x_1 x_1' + y_1 y_1') \right] \right\} \tag{3.22}$$

where Eq. (2.23) has been used, and constant phase factor $e^{ikn_0 d}$ has been omitted. Likewise, Eq. (3.22) can be rewritten as

$$K(x_1, y_1, x_1', y_1'; d_1', d, d_1) = \frac{k}{i2\pi B_s'} \exp\left\{ i\frac{kC_s'}{2A_s'}(x_1'^2 + y_1'^2) \right\} \exp\left\{ i\frac{kA_s'}{2B_s'}\left[\left(x_1 - \frac{x_1'}{A_s'} \right)^2 + \left(y_1 - \frac{y_1'}{A_s'} \right)^2 \right] \right\} \tag{3.23}$$

For $B_s' = 0$ or $B_s = n_0 B_s' = 0$, Eq. (3.23) yields

$$K(x_1, y_1, x_1', y_1'; d_1', d, d_1) = \frac{1}{A_s'} \exp\left\{ i\frac{kC_s'}{2A_s'}(x_1'^2 + y_1'^2) \right\} \delta\left(x_1 - \frac{x_1'}{A_s'}, y_1 - \frac{y_1'}{A_s'} \right) \tag{3.24}$$

Then, substitution of Eq. (3.24) into Eq. (3.20) provides

$$\psi_0(x_1', y_1') = \frac{1}{d_0 A_s'} \exp\left\{ i\frac{k}{2A_s'}\left(\frac{1}{d_0 A_s'} + C_s' \right)(x_1'^2 + y_1'^2) \right\} f_s\left(\frac{x_1'}{A_s'}, \frac{y_1'}{A_s'} \right) \tag{3.25}$$

Therefore, the output irradiance distribution is equal to the scaled replica of the transparency if the general condition

$$B_s' = 0 \tag{3.26}$$

is satisfied, where A_s' is the scaling factor or transverse magnification of the optical system.

From Eqs. (3.26) and (3.21) it follows that the distance from the output face of the GRIN lens to the image plane is given by

$$d_1' = -\frac{H_a(d)+n_0 d_1 H_f(d)}{n_0\left[\dot{H}_a(d)+n_0 d_1 \dot{H}_f(d)\right]} = -\left\{n_0 \frac{d}{dz}\ln[H_a(z)+n_0 d_1 H_f(z)]\Big|_{z=d}\right\}^{-1} \quad (3.27)$$

and that the transverse magnification is given by

$$m_t = H_f(d) + n_0 d_1' \dot{H}_f(d) = A_s' \quad (3.28)$$

Equation (3.28) can also be written as

$$m_t = [\dot{H}_a(d) + n_0 d_1 \dot{H}_f(d)]^{-1} = D_s'^{-1} \quad (3.29)$$

where Eqs. (3.27) and (3.21) have been used.

Equation (3.27) is called the GRIN lens law for imaging. It specifies, from a ray optics viewpoint, the particular distance where the rays emanating from a single object point will again cross in an image point.

Moreover, by taking into account Eqs. (3.28–3.29), Eq. (3.25) is expressed as

$$\psi_0(x_1', y_1') = \frac{1}{d_0 A_s'}\exp\left\{i\frac{k}{2A_s'}\left(\frac{D_s'}{d_0} + C_s'\right)(x_1'^2 + y_1'^2)\right\} f_s\left(\frac{x_1'}{A_s'}, \frac{y_1'}{A_s'}\right) \quad (3.30)$$

In short, imaging transformation for the optical system shown in Fig. 3.5 is described by the matrix

$$M_s = \begin{pmatrix} A_s' & 0 \\ C_s' + \frac{D_s'}{d_0} & D_s' \end{pmatrix} = \begin{pmatrix} A_s' & 0 \\ C_s' + \frac{D_s'}{d_0} & A_s'^{-1} \end{pmatrix} \quad (3.31)$$

Therefore, A_s' stands for the scaling factor, whereas $\left(C_s' + \frac{D_s'}{d_0}\right)A_s'^{-1}$ corresponds to the phase factor. For $C_s' + \frac{D_s'}{d_0} = 0$, we have

$$d_0 = -\frac{D_s'}{C_s'} = -\left[d_1 + \frac{\dot{H}_a(d)}{n_0 \dot{H}_f(d)}\right] \quad (3.32)$$

and the scaled image of the input complex amplitude transparency f_s without any additional quadratic phase factor is obtained. For illumination by a plane wave ($d_0 \to \infty$), the phase factor is canceled as $C_s = 0$ or $\dot{H}_f(d) = 0$, that is, as we use a telescopic GRIN lens.

Likewise, from Eq. (3.27) it is possible to find the front and back focal planes of the GRIN lens measured from the input and output faces of the lens, respectively. As $d_1 \to \infty$, we obtain the back working distance for uniform illumination

$$d_1' \to l' = -\frac{H_f(d)}{n_0 \dot{H}_f(d)} \quad (3.33)$$

which coincides with Eq. (3.14). As $d_1' \to \infty$, the front working distance is obtained

$$d_1 \to l = -\frac{\dot{H}_a(d)}{n_0 \dot{H}_f(d)} \tag{3.34}$$

Both equations indicate that, in general, the back and front working distances are different.

On the other hand, one of the most remarkable and useful properties of a lens is its inherent ability to perform bidimensional Fourier transforms. We will now study the general condition for which the output is given by the ordinary Fourier transform of the complex amplitude transparency $f_s(x_1, y_1)$.

Substituting Eq. (3.22) into Eq. (3.20), we have

$$\psi_0(x_1', y_1') = \frac{k}{i2\pi d_0 B_s'} \exp\left\{i\frac{kD_s'}{2B_s'}\left(x_1'^2 + y_1'^2\right)\right\} \cdot \int_{\Re^2} \exp\left\{i\frac{k}{2B_s'}\left[A_s' + \frac{B_s'}{d_0}\right]\left(x_1^2 + y_1^2\right)\right\} f_s(x_1, y_1) \exp\left\{-i\frac{k}{2B_s'}\left(x_1 x_1' + y_1 y_1'\right)\right\} dx_1 dy_1 \tag{3.35}$$

When $A_s' + \frac{B_s'}{d_0} = 0$ the output field is proportional to the ordinary Fourier transform of $f_s(x_1, y_1)$. Then, the general condition to obtain an ordinary Fourier transform can be written as $d_0 = -B_s'/A_s'$, that is

$$\begin{aligned} d_1' &= -\frac{H_a(d) + n_0(d_0 + d_1)H_f(d)}{n_0\left[\dot{H}_a(d) + n_0(d_0 + d_1)\dot{H}_f(d)\right]} \\ &= -\left\{n_0 \frac{d}{dz} \ln\left[H_a(z) + n_0(d_0 + d_1)H_f(z)\right]\Big|_{z=d}\right\}^{-1} \end{aligned} \tag{3.36}$$

where Eq. (3.21) has been used. Equation (3.36) represents the GRIN lens law for transforming and may be regarded as the imaging condition for the illumination source [3.6–3.7].

Thus, at a particular output plane for which d_1' satisfies Eq. (3.36), the complex amplitude distribution reduces to

$$\psi_0(x_1', y_1') = \frac{k}{i2\pi d_0 B_s'} \exp\left\{i\frac{kD_s'}{2B_s'}\left(x_1'^2 + y_1'^2\right)\right\} \cdot \int_{\Re^2} f_s(x_1, y_1) \exp\left\{-i2\pi\left(f_x x_1 + f_y y_1\right)\right\} dx_1 dy_1 \tag{3.37}$$

where

$$f_x = \frac{x'_1}{\lambda B'_s}; \quad f_y = \frac{y'_1}{\lambda B'_s} \tag{3.38}$$

are the spatial frequencies.

The quadratic phase factor that precedes the ordinary Fourier transform is canceled when $D'_s = 0$, that is

$$d_1 = -\frac{\dot{H}_a(d)}{n_0 \dot{H}_f(d)} = 1 \tag{3.39}$$

Thus, when the transparency is situated at the front focal plane of the GRIN lens, the ordinary Fourier transform of the transparency without any additional phase factor is achieved.

An important particular case of Eq. (3.36) occurs as $d_0 \to \infty$. In this case, the location of the transform plane is on the back focal plane

$$d'_1 = -\frac{H_f(d)}{n_0 \dot{H}_f(d)} = 1' \tag{3.40}$$

Equation (3.40) is satisfied for any position of the input plane when the transparency is uniformly illuminated by a plane wave.

Moreover, for uniform plane–wave illumination, the ordinary Fourier transform is obtained at the back focal plane without an additional phase factor when the transparency is situated at the front focal plane of the optical system,

$$\psi_0(x'_1, y'_1) = \frac{ikn_0 \dot{H}_f(d)}{2\pi d_0} \int_{\Re 2} f_s(x_1, y_1) \exp\{-i2\pi(f_x x_1 + f_y y_1)\} dx_1 dy_1 \tag{3.41}$$

where

$$f_x = -\frac{n_0 \dot{H}_f(d)}{\lambda} x'_1; \quad f_y = -\frac{n_0 \dot{H}_f(d)}{\lambda} y'_1 \tag{3.42}$$

are the spatial frequencies.

In short, ordinary Fourier transforming is described by the matrix

$$\begin{pmatrix} 0 & B'_s \\ C'_s + \dfrac{D'_s}{d_0} & D'_s \end{pmatrix} = \begin{pmatrix} 0 & -d_0 A'_s \\ (d_0 A'_s)^{-1} & D'_s \end{pmatrix} \tag{3.43}$$

B'_s indicates the scaling factor of the Fourier transform, and $D'_s B'^{-1}_s$ corresponds to the phase factor. For $D'_s = 0$, the phase factor is canceled.

In Eq. (3.43) it is easy to prove that

$$C'_s + \frac{D'_s}{d_0} = \frac{1}{d_0 A'_s} \tag{3.44}$$

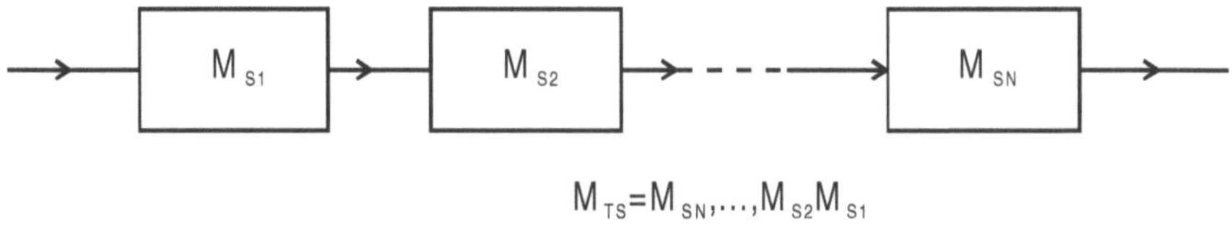

Fig. 3.6. Cascade of GRIN optical systems.

since

$$C'_s + \frac{D'_s}{d_0} = \frac{\dot{H}_a(d) + n_0(d_0 + d_1)\dot{H}_f(d)}{d_0} \tag{3.45}$$

and

$$\begin{aligned} d_0 A'_s &= d_0 \left[H_f(d) + n_0 d'_1 \dot{H}_f(d)\right] \\ &= d_0 \left[H_f(d) - \frac{H_a(d)\dot{H}_f(d) + n_0(d_0 + d_1)H_f(d)\dot{H}_f(d)}{\dot{H}_a(d) + n_0(d_0 + d_1)\dot{H}_f(d)}\right] \\ &= \frac{d_0}{\dot{H}_a(d) + n_0(d_0 + d_1)\dot{H}_f(d)} \end{aligned} \tag{3.46}$$

where Eqs. (1.79b), (3.21), and (3.36) have been used.

Finally, note that the impulse response of the optical system could be evaluated by the following integral

$$\begin{aligned} K(x_1, y_1, x'_1, y'_1; d_1, d, d'_1) = \int_{\Re^4} & K^{FP}(x_1, y_1, x_0, y_0; d_1) K^{GL}(x_0, y_0, x, y; d) \\ & K^{FP}(x, y, x'_1, y'_1; d'_1) dx dy dx_0 dy_0 \end{aligned} \tag{3.47}$$

where K^{FP} and K^{GL} are the impulse responses for free propagation through distances d_1 and d'_1 and for propagation along a GRIN lens of thickness d, respectively. In conclusion, the diffraction of a uniform wavefront of curvature radius d_0 on the complex amplitude transparency $f_s(x_1, y_1)$ in the optical system shown in Fig. 3.5 can be represented by the canonical integral transform, given by Eq. (3.20), parameterized by the matrix

$$\begin{pmatrix} A'_s + \dfrac{B'_s}{d_0} & B'_s \\ C'_s + \dfrac{D'_s}{d_0} & D'_s \end{pmatrix} \tag{3.48}$$

such that $B'_s = 0$ denotes the imaging condition with scaling factor A'_s, and $C'_s + \frac{D'_s}{d_0} = 0$ represents perfect scaled imaging; $A'_s + \frac{B'_s}{d_0} = 0$ indicates the ordinary Fourier transforming condition with scaling factor B'_s, and $D'_s = 0$ corresponds to perfect scaled transforming.

These results can be generalized for a cascade of GRIN optical systems (Fig. 3.6) whose ray-transfer matrices are $M_{S1}, M_{S2}, \ldots, M_{SN}$. The output of the first optical system operates as the input of the second system, and so on. Therefore the whole optical system is equivalent to a single GRIN optical system of ray-transfer matrix

$$M_{TS} = \prod_{j=1}^{N} M_{Sj} = M_{SN} \ldots M_{S2} M_{S1} = \begin{pmatrix} A'_{TS} & B'_{TS} \\ C'_{TS} & D'_{TS} \end{pmatrix} \tag{3.49}$$

whose kernel is given by

$$\begin{aligned} & K^{TS}\left(x_1, y_1, x'_N, y'_N; d_1, d^1, d'_1, d_2, d^2, d'_2, \ldots, d_N, d^N, d'_N\right) \\ & = \frac{k}{i2\pi B'_{TS}} \exp\left\{ i \frac{k}{2B'_{TS}} \left[\left(x'^2_N + y'^2_N\right) D'_{TS} + \left(x_1^2 + y_1^2\right) A'_{TS} - 2\left(x_1 x'_N + y_1 y'_N\right) \right] \right\} \end{aligned} \tag{3.50}$$

where $d^1, d^2, \ldots, d^N$ are the lengths of the N GRIN lenses.

3.4 Geometrical Optics of GRIN Lenses

In conventional lenses imaging is a result of discrete refraction occuring at the boundaries of homogeneous media of different refractive index. By using materials in which the refractive index varies in some controlled way, it is possible to form images by continuous refraction. Index variation may be axial, radial, or a combination of the two. GRIN lenses combine refraction at the end surfaces with continuous refraction within the lenses. In order to find object and image positions for GRIN lenses, it is convenient to define lens parameters in analogy with those of conventional homogeneous lenses. In this section we use the ray-transfer matrix to determine the cardinal elements of a single GRIN optical system (Fig. 3.7) with n_1 and n'_1 refractive indices of the object and image spaces, respectively. Knowledge of elements should be useful for designing GRIN lens systems. As a convention, we consider that in the object space the distances measured to the left are positives, while in the image space the distances measured to the right are positives (Fig. 3.7).

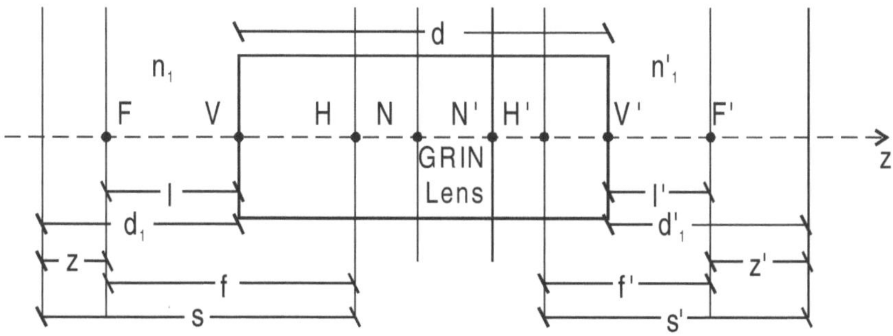

Fig. 3.7. Single GRIN optical system: Cardinal, object, and image planes, and distances between these planes and the lens faces.

The elements of the ray-transfer matrix of the GRIN optical system are given by [3.8]

$$A'_s = \frac{1}{n'_1}[n'_1 H_f(d) + n_0 d'_1 \dot{H}_f(d)] \tag{3.51a}$$

$$B'_s = \frac{1}{n_0}\left\{ n_1 H_a(d) + n_0 d_1 H_f(d) + \frac{n_0 d'_1}{n'_1}[n_1 \dot{H}_a(d) + n_0 d_1 \dot{H}_f(d)] \right\} \tag{3.51b}$$

$$C'_s = \frac{n_0}{n'_1}\dot{H}_f(d) \tag{3.51c}$$

$$D'_s = \frac{1}{n'_1}[n_1 \dot{H}_a(d) + n_0 d_1 \dot{H}_f(d)] \tag{3.51d}$$

with the requirement

$$A'_s D'_s - B'_s C'_s = \frac{n_1}{n'_1} \tag{3.52}$$

As previously mentioned, it is worthwhile to emphasize the quantitative information contained in the elements of the matrix. The image distance measured from the output face of the lens is given by $B'_s = 0$, that is

$$d'_1 = -\frac{n'_1[n_1 H_a(d) + n_0 d_1 H_f(d)]}{n_0[n_1 \dot{H}_a(d) + n_0 d_1 \dot{H}_f(d)]} = -\frac{n'_1}{n_0}\left\{ \frac{d}{dz} \ln[n_1 H_a(z) + n_0 d_1 H_f(z)] \Big|_{z=d} \right\}^{-1} \tag{3.53}$$

The linear or transverse magnification m_t is given by element A'_s

$$m_t = \frac{n'_1 H_f(d) + n_0 d'_1 \dot{H}_f(d)}{n'_1} \tag{3.54}$$

Using Eq. (3.53) we can rewrite this parameter as a function of the object distance measured from input face of the lens

$$m_t = \frac{n_1}{n_1 \dot{H}_a(d) + n_0 d_1 \dot{H}_f(d)} = \frac{n_1}{n_1'} D_s'^{-1} \tag{3.55}$$

The angular magnification m_a is obtained by element D_s'

$$m_a = \frac{n_1 \dot{H}_a(d) + n_0 d_1 \dot{H}_f(d)}{n_1'} \tag{3.56}$$

and using Eq. (3.53), we can express m_a as a function of the image distance

$$m_a = \frac{n_1}{n_1' H_f(d) + n_0 d_1' \dot{H}_f(d)} = \frac{n_1}{n_1'} A_s'^{-1} \tag{3.57}$$

From Eqs. (3.54–3.57) it follows that

$$m_a m_t = \frac{n_1}{n_1'} \tag{3.58}$$

The distance from the input face to the front focus F (front working distance) is given by $D_s' = 0$

$$VF = -\frac{n_1 \dot{H}_a(d)}{n_0 \dot{H}_f(d)} = l \tag{3.59}$$

The distance from the output face to the back focus F' (back working distance) is given by $A_s' = 0$

$$V'F' = -\frac{n_1' H_f(d)}{n_0 \dot{H}_f(d)} = l' \tag{3.60}$$

Equations. (3.59–3.60) coincide with Eqs. (3.33–3.34) for $n_1 = n_1' = 1$.

The principal planes have unit positive transverse magnification, therefore we can use the condition $m_t = 1$ to find them. The distance from the input face to the object principal plane can be obtained from Eq. (3.55), that is

$$VH = \frac{n_1[1 - \dot{H}_a(d)]}{n_0 \dot{H}_f(d)} \tag{3.61}$$

and the distance from the output face to the image principal plane is given by

$$V'H' = \frac{n_1'[1 - H_f(d)]}{n_0 \dot{H}_f(d)} \tag{3.62}$$

where Eq. (3.54) has been used.

Similarly, we can find the nodal planes by taking into account that the nodal planes have unit positive angular magnification. Therefore, the condition $m_a = 1$

is used to determine them. The distance from the input face to the object nodal plane follows from Eq. (3.56)

$$VN = \frac{n_1' - n_1 \dot{H}_a(d)}{n_0 \dot{H}_f(d)} \tag{3.63}$$

The distance from the output face to the image nodal plane is obtained from Eq. (3.57), that is

$$V'N' = \frac{n_1 - n_1' H_f(d)}{n_0 \dot{H}_f(d)} \tag{3.64}$$

On the other hand, to obtain the Gaussian formula for GRIN lenses, the distances from principal planes must be taken into account, that is

$$s = d_1 - VH = d_1 - \frac{n_1 [1 - \dot{H}_a(d)]}{n_0 \dot{H}_f(d)} \tag{3.65a}$$

$$s' = d_1' - V'H' = d_1' - \frac{n_1' [1 - H_f(d)]}{n_0 \dot{H}_f(d)} \tag{3.65b}$$

Then the front and focal distances are given, respectively, by

$$f = HF = VF - VH = -\frac{n_1}{n_0 \dot{H}_f(d)} \tag{3.66a}$$

$$f' = H'F' = V'F' - V'H' = -\frac{n_1'}{n_0 \dot{H}_f(d)} \tag{3.66b}$$

Note that for the zeros of $\dot{H}_f(d)$, the cardinal planes are at infinity, and the focal distances tend toward infinity.

From Eqs. (3.65–3.66) and (1.79b), it follows that Eq. (3.53) becomes the well–known Gaussian formula

$$\frac{n_1'}{s'} + \frac{n_1}{s} = \frac{n_1}{f} = \frac{n_1'}{f'} \tag{3.67}$$

In the same way, to obtain the Newtonian formula, we use the distances from the focal planes

$$z = d_1 - VF = d_1 + \frac{n_1 \dot{H}_a(d)}{n_0 \dot{H}_f(d)} \tag{3.68a}$$

$$z' = d_1' - V'F' = d_1' + \frac{n_1' H_f(d)}{n_0 \dot{H}_f(d)} \tag{3.68b}$$

Then using Eqs. (3.66) and (1.79b), the image distance (3.53) is easily rewritten into the Newtonian formula

$$zz' = ff' \tag{3.69}$$

Likewise, transverse and angular magnifications take the more familiar forms

$$m_t = -\frac{f}{z} = -\frac{z'}{f'} \tag{3.70a}$$

$$m_a = -\frac{n_1 z}{n_1' f} = -\frac{n_1 f'}{n_1' z'} \tag{3.70b}$$

where Eqs. (3.66) and (3.68) have been used.

Similarly, some expressions for the longitudinal magnification m_l can be obtained from Eqs. (3.67a), (3.69) and (3.70a), that is

$$m_l = \frac{dz}{dz'} = -\frac{f'f}{z^2} = -\frac{n_1' f^2}{n_1 z^2} = -\frac{n_1' z'^2}{n_1 f'^2} = -\frac{n_1'}{n_1} m_t^2 \tag{3.71}$$

Equation (3.71) can be rewritten as

$$m_l = -\frac{n_1 n_1'}{\left[n_0 d_1 \dot{H}_f(d) + n_1 \dot{H}_a(d)\right]^2} = -\frac{\left[n_1' H_f(d) + n_0 d_1' \dot{H}_f(d)\right]^2}{n_1 n_1'} \tag{3.72}$$

where Eqs. (3.54–3.55) have been used.

Moreover, from Eqs. (3.58) and (3.71) it follows that the relationship between magnifications is given by

$$m_t = -m_a m_l \tag{3.73}$$

Note that in a selfoc lens if $n_1 = n_1'$, object and image principal and nodal planes coincide respectively, and that $f = f'$ and $l = l'$ for any lens length since $\dot{H}_a = H_f$ from Eqs. (1.96).

As an example, we show in Figs. 3.8 and 3.9 the equi-index surfaces and the behavior of the cardinal planes of an exponential GRIN lens surrounded by free space as the lens length is varied. The gradient parameter is given by

$$g(z) = g_0 \exp\left\{-z/L\right\} \tag{3.74}$$

where $g_0 = g(0)$, and L is a parameter.

The equi-index surfaces can be expressed as

$$r(z) = r_0 \exp\left\{z/L\right\} \tag{3.75}$$

where

$$r_0 = \frac{\sqrt{n_0^2 - n^2(x, y, 0)}}{n_0 g_0} \tag{3.76}$$

and Eq. (3.1) has been used.

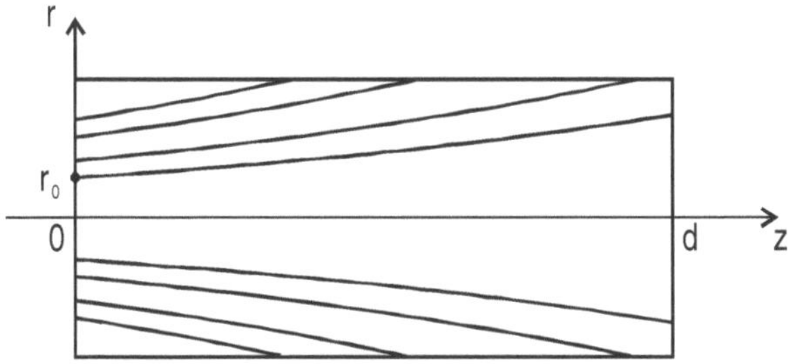

Fig. 3.8. Equi-index surfaces of an exponential GRIN lens. Calculations have been made for $n_0 = 1.5$, $g_0 = 0.3\,\text{mm}^{-1}$, $L = 25\,\text{mm}$, and $d = 50\,\text{mm}$.

For this GRIN lens, we have [3.8]

$$H_a(d) = \frac{\pi L}{2}\{-Y_0[g_0L]J_0[g(d)L] + J_0[g_0L]Y_0[g(d)L]\} \tag{3.77a}$$

$$H_f(d) = \frac{\pi g_0 L}{2}\{-Y_1[g_0L]J_0[g(d)L] + J_1[g_0L]Y_0[g(d)L]\} \tag{3.77b}$$

$$\dot{H}_a(d) = \frac{\pi g(d)L}{2}\{-Y_0[g_0L]J_1[g(d)L] + J_0[g_0L]Y_1[g(d)L]\} \tag{3.77c}$$

$$\dot{H}_f(d) = \frac{\pi g_0 g(d)L}{2}\{-Y_1[g_0L]J_1[g(d)L] + J_1[g_0L]Y_1[g(d)L]\} \tag{3.77d}$$

where J_q and Y_q are the q-order Bessel functions of the first and second kind, respectively.

Equations (3.59–3.64) become

$$VF = \frac{Y_0[g_0L]J_1[g(d)L] - J_0[g_0L]Y_1[g(d)L]}{n_0 g_0\{J_1[g_0L]Y_1[g(d)L] - Y_1[g_0L]J_1[g(d)L]\}} \tag{3.78a}$$

$$V'F' = \frac{Y_1[g_0L]J_0[g(d)L] - J_1[g_0L]Y_0[g(d)L]}{n_0 g(d)\{J_1[g_0L]Y_1[g(d)L] - Y_1[g_0L]J_1[g(d)L]\}} \tag{3.78b}$$

$$VH = VN = \frac{2 + \pi g(d)L\{Y_0[g_0L]J_1[g(d)L] - J_0[g_0L]Y_1[g(d)L]\}}{\pi n_0 g_0 g(d)L\{J_1[g_0L]Y_1[g(d)L] - Y_1[g_0L]J_1[g(d)L]\}} \tag{3.78c}$$

$$V'H' = V'N' = \frac{2 + \pi g_0 L\{Y_1[g_0L]J_0[g(d)L] - J_1[g_0L]Y_0[g(d)L]\}}{\pi n_0 g_0 g(d)L\{J_1[g_0L]Y_1[g(d)L] - Y_1[g_0L]J_1[g(d)L]\}} \tag{3.78d}$$

Figure 3.9 depicts variation with exponential GRIN lens length of (a) the distances from the lens faces to the focal planes, and (b) the distances from the lens faces to the principal (nodal) planes up to the lens length corresponding to the

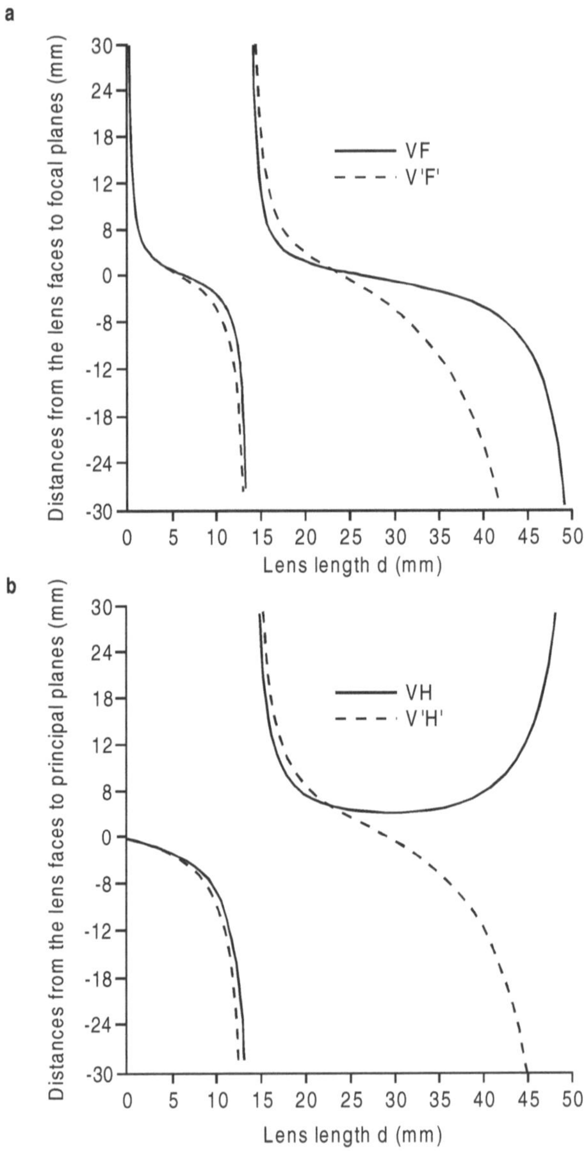

Fig. 3.9. Variation with exponential GRIN lens length of **a** the distances from the lens faces to the focal planes, **b** the distances from the lens faces to the principal (nodal) planes. Calculations have been made for lens parameters $n_0 = 1.5$, $g_0 = 0.3\,mm^{-1}$, $L = 25\,mm$ and object and image space refractive indices $n_1 = n_1' = 1$

second zero of $\dot{H}_f(d)$. Near the first and second zeros of $\dot{H}_f(d)$, obtained for lens lengths of approximately 14 and 49 mm the distances go to infinity as expected. Likewise, in general the distances are different, and they coincide for small lengths only. In particular, behavior of distances VH and V′H′ are very different when the length tends toward the second zero of $\dot{H}_f(d)$.

The broadening of the graphs toward the right is a logical effect of the gradient parameter. It makes the transverse gradient smaller, subsequently, the GRIN lens is more homogeneous. The period of the rays inside the lens increases due to the divergence of equi-index surfaces, and therefore a greater length of lens is required to obtain the same effects.

3.5 Effective Radius, Numerical Aperture, Aperture Stop and Pupils

It is well–known that for an optical system it is not enough to be able to predict the image position and its size, other important properties of an image include, for instance, the brightness and size of the field of view. These topics require a study of the limitations of the spatial and angular extents of the light beams by stops and apertures. Even if the optical system under consideration accepts light rays that are not paraxial, it is usually sufficient to treat stops and apertures and related effects by using methods of paraxial optics [3.9–3.21].

In this section we will study, under paraxial approximation, these topics for GRIN lenses with rotational symmetry surrounded by free space. The paraxial ray path through a GRIN lens is given by Eq. (1.106), that is

$$r(z) = R(z)\sin\left[\int_0^z g(z')\,dz' + \delta_0\right] \tag{3.79}$$

In a GRIN lens of radius a and thickness d a ray will be confined if its height satisfies the condition

$$\frac{\sqrt{\dot{r}_0^2 + g_0^2 r_0^2}}{[g_0 g(z)]^{1/2}} \le a \qquad \text{for } 0 \le z \le d \tag{3.80}$$

where Eq. (1.108a) has been used.

Equation (3.80) means that not all rays reaching the input face of the lens will be confined through it and that there is an angular limitation in the cone of light entering the lens (Fig. 3.10).

Equation (3.80) for the paraxial marginal ray becomes

$$\dot{r}_m^2 + (g_0 a_e)^2 = a^2 g_0 g(z) \tag{3.81}$$

where a_e and $\dot{r}_m$ are the position and slope of this ray at the input face.

The marginal ray starting from an off-axis object point at a distance d_1 from the input face with position b and slope $\dot{r}_{1m}$ verifies that

$$\left(\frac{\dot{r}_{1m}}{n_0}\right)^2 + (g_0 a_e)^2 = a^2 g_0 g(z) \tag{3.82}$$

where

$$a_e = b + d_1 \dot{r}_{1m} \tag{3.83}$$

is the position of the marginal ray on the input face of the GRIN lens.

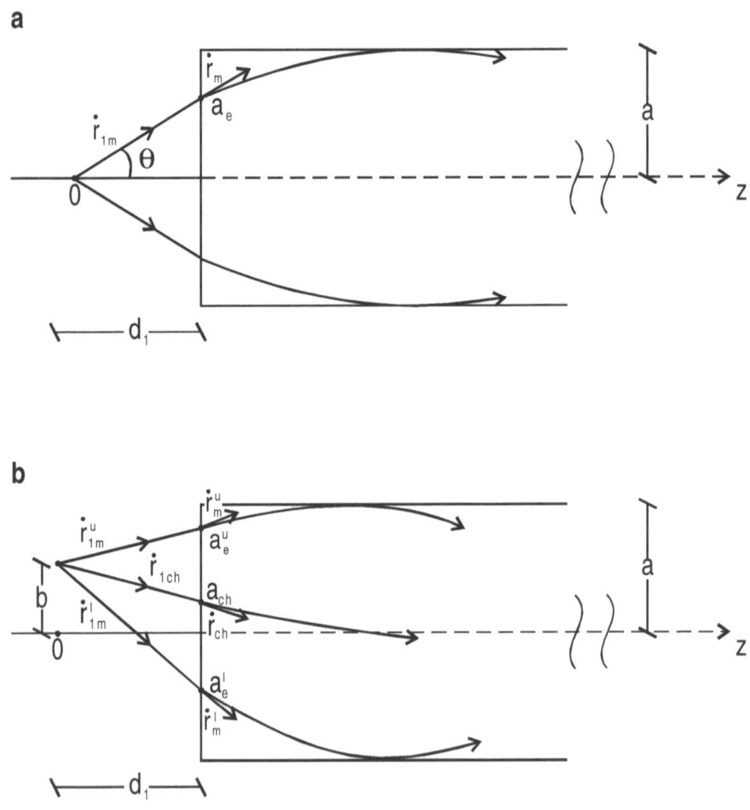

Fig. 3.10. Limitation of rays in a GRIN lens for **a** an on-axis object point, **b** an off-axis object point. Subindices m and ch denote marginal and chief rays, and superscripts u and l indicate upper and lower marginal rays, respectively.

From Eqs. (3.82–3.83) it follows that

$$\dot{r}_{1m}^{\binom{u}{l}} = \frac{-n_0^2 g_0^2 b d_1 \pm n_0 \sqrt{[1+n_0^2 g_0^2 d_1^2]a^2 g_0 g(z) - g_0^2 b^2}}{1+n_0^2 g_0^2 d_1^2} \tag{3.84}$$

with superscript u for plus and l for minus since each object point has upper and lower marginal rays and meridional rays corresponding to the solutions in Eq. (3.84).

Likewise, the chief ray is the ray in the center of the cone of light from any point of the object. Equations (3.83–3.84) show that the slope of the chief ray is

$$\dot{r}_{1ch} = -\frac{n_0^2 g_0^2 d_1 b}{1+n_0^2 g_0^2 d_1^2} \tag{3.85a}$$

and that its position on the input face of the GRIN lens is

$$a_{ch} = \frac{b}{1+n_0^2 g_0^2 d_1^2} \tag{3.85b}$$

The trajectory of the chief ray along a GRIN lens can be expressed as

$$r_{ch}(z) = H_f(z)a_{ch} + H_a(z)\dot{r}_{ch} = \frac{b}{1+n_0^2 g_0^2 d_1^2}[H_f(z) - n_0 g_0^2 d_1 H_a(z)] \tag{3.86a}$$

$$\dot{r}_{ch}(z) = \dot{H}_f(z)a_{ch} + \dot{H}_a(z)\dot{r}_{ch} = \frac{b}{1+n_0^2 g_0^2 d_1^2}[\dot{H}_f(z) - n_0 g_0^2 d_1 \dot{H}_a(z)] \tag{3.86b}$$

where $\dot{r}_{ch} = \dot{r}_{1ch}/n_0$ and Eqs. (1.77) and (3.85) have been used.

Moreover, from Eqs. (3.83–3.84) we can find the positions of the upper and lower marginal rays on the input face

$$a_e^{\binom{u}{l}} = \frac{b \pm n_0 d_1 \sqrt{[1+n_0^2 g_0^2 d_1^2]a^2 g_0 g(z) - g_0^2 b^2}}{1+n_0^2 g_0^2 d_1^2} \tag{3.87a}$$

which determine the effective aperture of the GRIN lens given by

$$\text{E.A.} = \frac{2n_0 d_1 \sqrt{[1+n_0^2 g_0^2 d_1^2]a^2 g_0 g(z) - g_0^2 b^2}}{1+n_0^2 g_0^2 d_1^2} \tag{3.87b}$$

The last equation indicates that effective aperture decreases as b increases. Equation (3.87a) can be written in terms of the chief ray as

$$a_e^{\binom{u}{l}} = a_{ch}\left[1 \pm n_0 g_0 d_1 \sqrt{\frac{a^2 g(z)}{g_0 b a_{ch}} - 1}\right] \tag{3.88}$$

where Eqs. (3.84) and (3.85b) have been used.

For the on-axis object point, b = 0 and the above equations reduce to

$$\dot{r}_{1m} = \pm \frac{n_0 a [g_0 g(z)]^{1/2}}{\sqrt{1 + n_0^2 g_0^2 d_1^2}} \tag{3.89a}$$

$$a_{ch} = 0 \tag{3.89b}$$

$$a_e = d_1 \dot{r}_{1m} = \pm \frac{n_0 d_1 a [g_0 g(z)]^{1/2}}{\sqrt{1 + n_0^2 g_0^2 d_1^2}} a \tag{3.89c}$$

From Eqs. (3.89a–b) it follows that the marginal rays generate a revolution light cone around the z–axis subtending at O an angle θ (Fig. 3.10a). The input numerical aperture, a quantity specifying the light-gathering power of an optical system, is given by

$$NA_{in} = \sin\theta = \sqrt{\frac{a_e^2}{a_e^2 + d_1^2}} = \frac{n_0 a [g_0 g(z)]^{1/2}}{\sqrt{1 + [a^2 g(z) + d_1^2 g_0] n_0^2 g_0}} \tag{3.90}$$

Equation (3.89c) defines the effective radius of the input face of the GRIN lens for which the rays are confined inside it. Therefore the evolution of the position and slope of the marginal rays can be written as

$$r_m(z) = a_e F(z) \tag{3.91a}$$

$$\dot{r}_m(z) = a_e \dot{F}(z) \tag{3.91b}$$

where

$$\overset{(\cdot)}{F}(z) = \overset{(\cdot)}{H}_f(z) + \frac{\overset{(\cdot)}{H}_a(z)}{n_0 d_1} \tag{3.92}$$

In this way, the maximum ray deviation from the center axis is smaller than or equal to the edge of the GRIN lens, and the transmittance function given by Eq. (3.7) must be redefined as follows

$$t_e(r_0, r; d) = t(r_0, r; d)\,\mathrm{cir}\left\{\frac{r}{a_e |F(d)|}\right\} \tag{3.93}$$

where t_e is the effective transmittance function of the GRIN lens, and cir is the circle function representing a circle of radius $a_e|F(d)|$ on the output face of the lens. In other words, the transmittance function t, when we have neglected the finite extent of the GRIN lens radius, is modulated by a circular exit aperture of effective radius $a_e|F(d)|$.

Of particular importance is the case where the object is located at infinite distance. In this case $d_1 \to \infty$, and the input effective radius coincides with the physical radius of the GRIN lens. The effective transmittance function reduces to

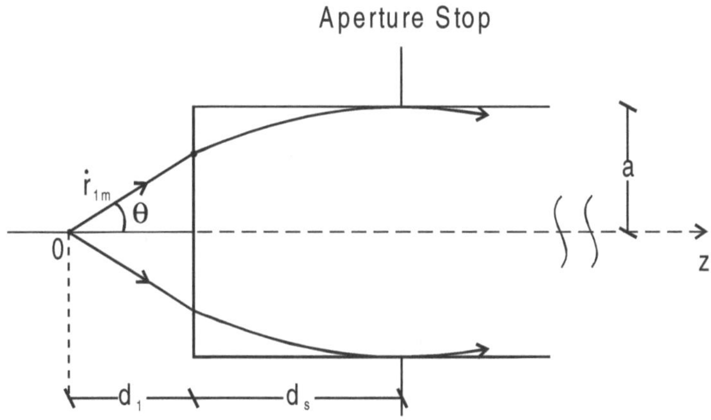

Fig. 3.11. Aperture stop in a GRIN lens for an on-axis object point.

$$t_e(r;d) = t(r;d)\mathrm{cir}\left\{\frac{r}{a|H_f(d)|}\right\} \tag{3.94}$$

where t(r;d) is given by Eq. (3.8).

On the other hand, the aperture stop is the rim of the lens or the stop that physically limits the cone of rays, from the axial object point, passing through an optical system. For the aperture stop in a GRIN lens, the position and the slope of a marginal ray are given by

$$r_m(d_s) = a \tag{3.95a}$$

$$\dot{r}_m(d_s) = 0 \tag{3.95b}$$

where d_s is the position of the aperture stop measured from the input face of the lens (Fig. 3.11).

From Eqs. (3.91–3.92) and (1.79b) it follows that the above equations can be written as

$$a_e = a\dot{H}_a(d_s) \tag{3.96a}$$

$$\frac{\dot{H}_f(d_s)}{\dot{H}_a(d_s)} = -\frac{1}{n_0 d_1} \tag{3.96b}$$

Equation (3.96a) indicates that the effective radius is proportional to the slope of the axial ray at d_s, and Eq. (3.96b) provides the aperture stop position, that is

$$\tan\left[\int_0^{d_s} g(z')\,dz'\right] = \frac{1}{n_0 g_0 d_1} \tag{3.97}$$

where Eqs. (1.95) have been used.

Once the position of the aperture stop is known, the entrance and exit pupils can be evaluated. The image of the aperture stop formed by the part of the lens between the aperture stop and the on-axis object point O will provide the entrance pupil, so, by taking into account Eqs. (3.27) and (3.29) for an object placed on the input face of a GRIN lens of thickness d_s, the position of the entrance pupil measured from the front face of the GRIN lens can be evaluated

$$d_{enp} = -\frac{H_a(d_s)}{n_0 \dot{H}_a(d_s)} = \frac{1}{n_0^2 g_0 g(d_s) d_1} \tag{3.98}$$

and the entrance pupil radius can be found

$$a_{enp} = \frac{a}{\dot{H}_a(d_s)} = \frac{a g_0^{1/2}}{g^{1/2}(d_s)\cos\left[\int_0^{d_s} g(z')\,dz'\right]} \tag{3.99}$$

where Eqs. (1.92a), (1.95b), and (3.97) have been used. Equation (3.99) from trigonometric relationships and Eqs. (3.89c) and (3.96a) can be expressed as

$$a_{enp} = \frac{n_0 g_0^{3/2} d_1 a}{g^{1/2}(d_s)\sqrt{1+n_0^2 g_0^2 d_1^2}} = \frac{g_0}{g(d_s)} a_e \tag{3.100}$$

The image of the aperture stop formed by the part of the lens between the aperture stop and the image O′ of the axial object point located at d_1' from the output face of the GRIN lens will provide the exit pupil. The position of the exit pupil measured from the output face of the lens is given by

$$d'_{exp} = -\frac{H_a(d_s - d)}{n_0 \dot{H}_a(d_s - d)} = \frac{1}{n_0^2 g(d) g(d_s) d_1'} \tag{3.101}$$

and its size

$$a'_{exp} = \frac{a}{\dot{H}_a(d_s - d)} = \frac{a g^{1/2}(d)}{g^{1/2}(d_s)\cos\left[\int_{d_s}^{d} g(z')\,dz'\right]} = \frac{g(d)}{g(d_s)} a_e' \tag{3.102}$$

where $a_e' = a_e|F(d)|$ is the output effective radius of the GRIN lens.

Note that Eqs. (3.101–3.102) can be obtained from Eqs. (3.98–3.100) replacing g_0, d_1 and a_e by $g(d)$, d_1', and a_e', respectively.

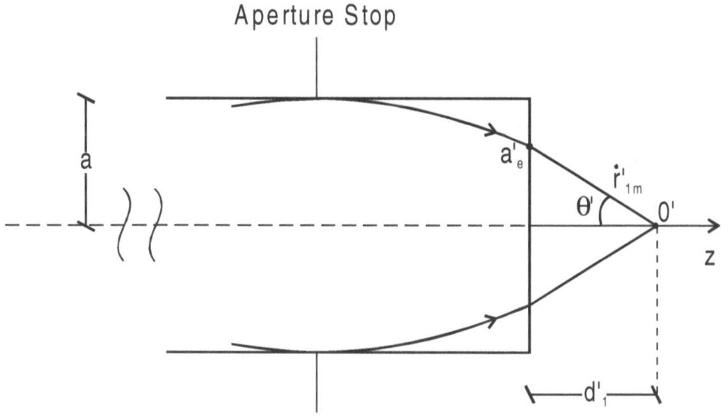

Fig. 3.12. Limitation of rays in the image space.

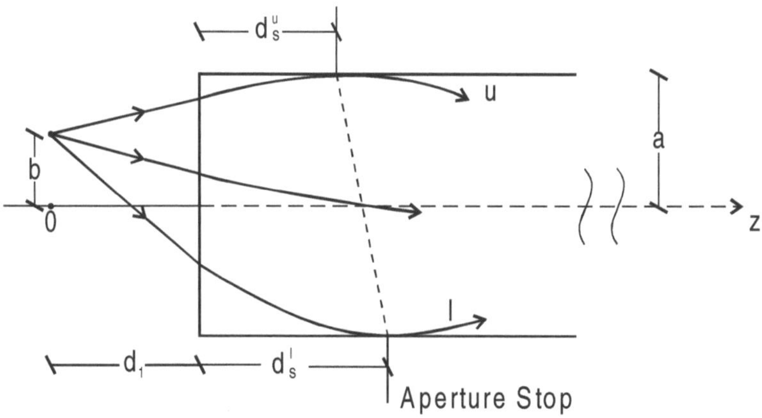

Fig. 3.13. Aperture stop for an off-axis object point.

Likewise, there are other conjugated topics related to the image space (Fig. 3.12). They are the exiting angle of the marginal rays, which determines the output numerical aperture, and, as mentioned above, the output effective radius. The output numerical aperture is given by

$$NA_{out} = \sin\theta' = \frac{n_0 a[g(d)g(z)]^{1/2}}{\sqrt{1+[a^2 g(z)+d_1'^2 g(d)]n_0^2 g(d)}} = \sqrt{\frac{a_e^2}{a_e^2+d_1'^2}} \qquad (3.103)$$

Input and output numerical apertures are important parameters when we use a GRIN lens as a collimator or focuser in optical connections to evaluate the collimated beam size or the focused spot size.

We return now to off-axis objects. As shown in Fig. 3.13, the limitation in the cone of light of an off-axis object point will happen in the plane named d_s^u for the upper meridional ray and in plane d_s^l for the lower one. The turning position of the upper meridional ray shifts toward the input face of the lens, and for the lower one, it moves away from the input face. It acts like an aperture stop that tilts, and there will be a gradual loss of light until no image is transmitted. This causes vignetting that completely blocks rays for off-axis points larger than a threshold height of the object.

Extreme positions of the aperture stop for the upper and lower marginal rays can be evaluated from the following conditions written, in compact form, as

$$r_m\left[d_s^{\binom{u}{l}}\right] = \pm a \tag{3.104a}$$

$$\dot{r}_m\left[d_s^{\binom{u}{l}}\right] = 0 \tag{3.104b}$$

After a straightforward calculation, we have

$$a_e^{\binom{u}{l}} = \pm a\dot{H}_a\left[d_s^{\binom{u}{l}}\right] \tag{3.105}$$

Substituting Eq. (1.95a) into Eq. (3.105), the extreme positions of the aperture stop are given by

$$\int_0^{d_s^{\binom{u}{l}}} g(z')\,dz' = \cos^{-1}\left\{\frac{g_0^{1/2} a_e^{\binom{u}{l}}}{a g^{1/2}\left[d_s^{\binom{u}{l}}\right]}\right\} \tag{3.106}$$

Moreover, the position of the center of the tilted aperture stop is the position d_s^c on the optical axis where the chief ray crosses it. For the center position, $r_{ch}(d_s^c) = 0$, and Eq. (3.86a) reduces to

$$\int_0^{d_s^c} g(z')\,dz' = \frac{1}{n_0 g_0 d_1} \tag{3.107}$$

which coincides with the position of the aperture stop for an on-axis object point.

There is a maximum height b_m of the object for which the GRIN lens will accept any ray. This is when b is large enough to cancel effective aperture. From Eq. (3.87b) it follows that

$$b_m = a\sqrt{\frac{(1+n_0^2 g_0^2 d_1^2)g(z)}{g_0}} \tag{3.108}$$

Therefore b_m defines the spatial extent of the object, that is, the size of the field of view in the object space. For the image space, the size of the field of view will be given by

$$b'_m = b_m \left[H_f(d) + n_0 d'_1 \dot{H}_f(d)\right] \tag{3.109}$$

where Eq. (3.54) has been used, and d'_1 is the image distance measured from the output face of the lens.

3.6
Diffraction-Limited Propagation of Light in a GRIN Lens

In order to show the diffraction-limited propagation in a GRIN lens of radius a and thickness d illuminated by a uniform wavefront, we now consider the lens being illuminated by a monochromatic uniform spherical wave of curvature radius d_1 and wavelength λ (Fig. 3.14). The complex amplitude distribution at some plane z>0 within the GRIN lens is given by [3.22–3.23]

$$\psi(r;z) = \int_0^{2\pi}\int_0^{a_s} \psi(r_0;0) K(r_0, r, \Omega_0, \Omega; z)\, r_0 dr_0 d\Omega_0 \tag{3.110}$$

where K is the impulse response of the lens given by Eq. (2.20) in polar coordinates

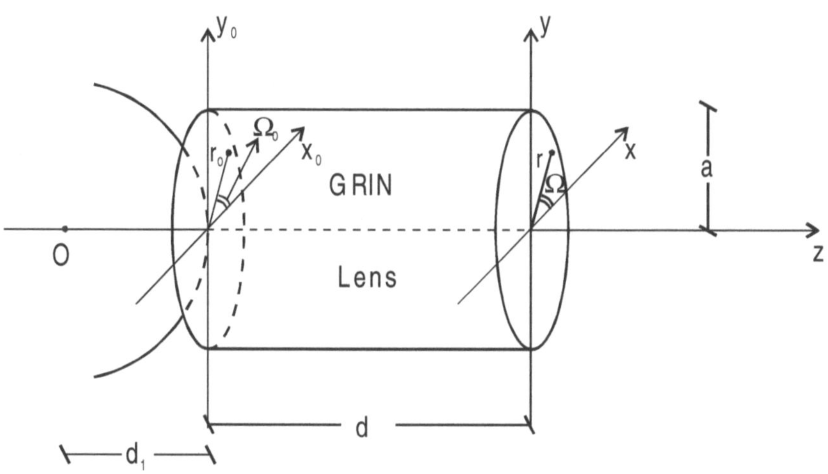

Fig. 3.14. Illustrating the diffraction-limited propagation in a GRIN lens.

$$K(r_0, r, \Omega_0, \Omega; z) = \frac{kn_0}{i2\pi H_a(z)} \exp\{ikn_0 z\}$$
$$\exp\left\{i\frac{kn_0}{2H_a(z)}\left[r^2\dot{H}_a(z) + r_0^2 H_f(z) - 2rr_0\cos(\Omega_0 - \Omega)\right]\right\} \tag{3.111}$$

where

$$x_{(0)} = r_{(0)} \cos\Omega_{(0)} \tag{3.112a}$$

$$y_{(0)} = r_{(0)} \sin\Omega_{(0)} \tag{3.112b}$$

the coordinates transformation,

$$\psi(r_0; 0) = \frac{1}{d_1}\exp\left\{i\frac{\pi}{\lambda d_1} r_0^2\right\} \tag{3.113}$$

the complex amplitude distribution at the input face of the lens, and a_e the input effective radius for spherical illumination.

Substituting Eqs. (3.111) and (3.113) into Eq. (3.110) and introducing dimensionless variables

$$u(z) = \frac{kn_0 a_e^2}{H_a(z)} F(z) \tag{3.114a}$$

$$v(z) = \frac{kn_0 r a_e}{H_a(z)} \tag{3.114b}$$

$$w = \frac{r_0}{a_e} \tag{3.114c}$$

Where Eq. (3.92) has been used, Eq. (3.110) may be rewritten as

$$\psi(r; z) = \frac{u(z)}{id_1 F(z)} \exp\{ikn_0 z\} \exp\left\{i\frac{u(z)\dot{H}_a(z)}{2a_e^2 F(z)} r^2\right\}.$$
$$\int_0^1 J_0[v(z)w] \exp\left\{i\frac{u(z)w^2}{2}\right\} w\,dw \tag{3.115}$$

where

$$J_0(vw) = \frac{1}{2\pi}\int_0^{2\pi} \exp\{-ivw\cos(\Omega_0 - \Omega)\}\, d\Omega_0 \tag{3.116}$$

is the zero-order Bessel function of the first kind.

On the other hand, Eqs. (3.114a–b) provide the relationship between u and v,

$$\frac{u(z)}{v(z)} = \frac{a_e F(z)}{r} \tag{3.117}$$

From the geometrical optics viewpoint, the numerator of Eq. (3.117) represents the trajectory of the marginal rays through the GRIN lens given by Eq. (3.91a). Then, the relationship between u and v defines, at any plane z inside the lens, the geometrically illuminated region as $\left|\frac{u}{v}\right\rangle 1$, the geometrical shadow as $\left|\frac{u}{v}\right\langle 1$, and the edge of the geometrical shadow for $\left|\frac{u}{v}\right| = 1$.

The earlier diffraction integral can be evaluated in terms of the Lommel functions [3.24]

$$\psi(r;z) = \frac{u(z)}{i2d_1F(z)}\exp\{ikn_0z\}\exp\left\{i\frac{u(z)\dot{H}_a(z)}{2a_e^2F(z)}r^2\right\}[C(u,v)+iS(u,v)] \quad (3.118)$$

where

$$C+iS = \frac{2}{u}\exp\left\{i\frac{u}{2}\right\}\left[i\exp\left\{-i\frac{(u^2+v^2)}{2u}\right\} - [iV_0(u,v)+V_1(u,v)]\right] \text{ for } \left|\frac{u}{v}\right\rangle 1 \quad (3.119a)$$

$$C+iS = \frac{i\exp\left\{i\frac{u}{2}\right\}}{u}[\exp\{-iu\}-J_0(u)] \text{ for } \left|\frac{u}{v}\right| = 1 \quad (3.119b)$$

and

$$C+iS = \frac{2}{u}\exp\left\{i\frac{u}{2}\right\}[U_1(u,v)-iU_2(u,v)] \text{ for } \left|\frac{u}{v}\right\langle 1 \quad (3.119c)$$

where U_1, U_2, V_0, and V_1 are the Lommel functions

$$U_n(u,v) = \sum_{s=0}^{\infty}(-1)^s\left(\frac{u}{v}\right)^{n+2s} J_{n+2s}(v) \quad (3.120a)$$

$$V_n(u,v) = \sum_{s=0}^{\infty}(-1)^s\left(\frac{v}{u}\right)^{n+2s} J_{n+2s}(v) \quad (3.120b)$$

Inserting Eqs. (3.119) into Eq. (3.118), taking into account Eq. (1.79b), and setting z=d, the complex amplitude distributions on the different geometrical regions at the output face of the lens are obtained.

For the geometrically illuminated region, $|u\rangle|v|$, and the complex field is given by

$$\psi(r;d) = \frac{1}{d_1F(d)}\exp\{ikn_0d\}\left\{\exp\left[i\frac{kn_0\dot{F}(d)}{2F(d)}r^2\right]\right.$$
$$\left. +\exp\left\{i\frac{u(d)}{2}\left[\frac{\dot{H}_a(d)r^2}{a_ea_e'(d)}+1\right]\right\}[iV_1(u,v)-V_0(u,v)]\right\} \quad (3.121)$$

where $a'_e(d) = a_e|F(d)|$ is the effective radius at the output face of the lens, and $\dot{F}(d)$ is the derivative of F(z) with respect to z evaluated at z = d.

The first term in Eq. (3.121) is the complex amplitude distribution at z = d for diffraction–free propagation through the lens, and the second one is the field due to diffraction–limited propagation.

For the edge of the geometrical shadow $|u| = |v|$, and we have

$$\psi(r;d) = \frac{1}{2d_1F(d)}\exp\{ikn_0d\}\left\{\exp\left[i\frac{kn_0\dot{F}(d)}{2F(d)}r^2\right] - J_0(u)\exp\left\{i\frac{u(d)}{2}\left[\frac{\dot{H}_a(d)r}{a_e}+1\right]\right\}\right\} \tag{3.122}$$

The resulting complex amplitude distribution may be regarded as the sum of the half–complex amplitude distribution for diffraction–free propagation and another term due to diffractional effects of the finite extent of the lens.

For the geometrical shadow $|u|\langle|v|$, and, in this case, the complex amplitude distribution is expressed as

$$\psi(r;d) = -\frac{1}{d_1F(d)}\exp\{ikn_0d\}\exp\left\{i\frac{u(d)}{2}\left[\frac{\dot{H}_a(d)r^2}{a_ea'_e(d)}+1\right]\right\}\cdot\left[iU_1(u,v)+U_2(u,v)\right] \tag{3.123}$$

Thus for the geometrical shadow the complex amplitude distribution at z=d can be regarded as being only due to diffraction–limited propagation.

From Eqs. (3.121–3.123) we obtain the irradiance distribution at the output face of the GRIN lens

$$I(r;d) = I_0(d)\left[1+V_0^2(u,v)+V_1^2(u,v)-2V_0(u,v)\cos\left\{\frac{kn_0}{2H_a(d)F(d)}[a_e'^2(d)+r^2]\right\} - 2V_1(u,v)\sin\left\{\frac{kn_0}{2H_a(d)F(d)}[a_e'^2(d)+r^2]\right\}\right] \tag{3.124a}$$

for $|u|\rangle|v|$,

$$I(r;d) = \frac{I_0(d)}{4}\left[1+J_0^2(u)-2J_0(u)\cos\left\{\frac{kn_0r^2}{H_a(d)F(d)}\right\}\right] \tag{3.124b}$$

for $|u| = |v|$, and

$$I(r;d) = I_0(d)[U_1^2(u,v)+U_2^2(u,v)] \tag{3.124c}$$

for $|u|\langle|v|$, where

$$I_0(d) = \frac{1}{d_1^2F^2(d)} \tag{3.125}$$

is the irradiance distribution for diffraction-free propagation at the output face of the GRIN lens.

We now consider a special case of interest. For points on axis, v = 0 and Eq. (3.115) can be integrated exactly to give

$$\psi(0;z) = \frac{1}{d_1 F(z)} \exp\{ikn_0 z\}\left\{1 - \exp\left[i\frac{u(z)}{2}\right]\right\} \tag{3.126}$$

Hence the irradiance along the axis of the GRIN lens is expressed as

$$I(0,z) = \frac{k^2 n_0^2 a_e^4}{4d_1^2 H_a^2(z)} \operatorname{sinc}^2\left[\frac{u(z)}{4}\right] \tag{3.127}$$

The zeros of irradiance are given by

$$\frac{kn_0 a_e^2 F(z_q)}{4H_a(z_q)} = q\pi \quad (q = \pm 1; \pm 2; ...) \tag{3.128}$$

Equation (3.128) can be written as

$$\frac{H_f(z_q)}{H_a(z_q)} = \frac{2q\lambda}{n_0 a_e^2} - \frac{1}{n_0 d_1} \tag{3.129}$$

where Eq. (3.92) has been used.

Equation (3.129) provides positions z_q^{oa} on the axis for which the irradiance cancels, that is

$$\cot\left[\int_0^{z_q^{oa}} g(z')dz'\right] = \frac{2q\lambda}{n_0 g_0 a_e^2} - \frac{1}{n_0 d_1 g_0} \tag{3.130}$$

where Eqs. (1.92) have been used.

Likewise, of particular importance is the case when the GRIN lens is illuminated by a uniform plane wave ($d_1 \to \infty$). In this case, the above equations can be replaced by simpler equations, and hence an easier study of light propagation can be made. For plane illumination, the effective radius reduces to the physical radius of the GRIN lens, and Eqs. (3.124) become

$$I(r;d) = \left[1 + V_0^2(u,v) + V_1^2(u,v) - 2V_0(u,v)\cos\left\{\frac{kn_0}{2H_a(d)H_f(d)}[a_e'^2(d) + r^2]\right\}\right.$$
$$\left. - 2V_1(u,v)\sin\left\{\frac{kn_0}{2H_a(d)H_f(d)}[a_e'^2(d) + r^2]\right\}\right]\frac{1}{H_f^2(d)} \tag{3.131a}$$

for the geometrically illuminated region,

$$I(r;d) = \left[1 + J_0^2(u) - 2J_0(u)\cos\left\{\frac{kn_0 r^2}{H_a(d)H_f(d)}\right\}\right]\frac{1}{4H_f^2(d)} \tag{3.131b}$$

for the edge of geometrical shadow, and

$$I(r;d) = [U_1^2(u,v) + U_2^2(u,v)]\frac{1}{H_f^2(d)} \tag{3.131c}$$

for the geometrical shadow, provided that the plane wave has unit amplitude and where

$$u(d) = \frac{kn_0 a^2 H_f(d)}{H_a(d)} \tag{3.132a}$$

$$v(d) = \frac{kn_0 ar}{H_a(d)} \tag{3.132b}$$

$$a'_e(d) = a|H_f(d)| \tag{3.132c}$$

We now study light propagation through the GRIN lens. For image planes z_m such that $H_a(z_m) = 0$, Eq. (3.111) becomes

$$K(r_0, r, \Omega_0, \Omega; z_m) = \frac{\exp\{ikz_m\}}{r_0 H_f(z_m)} \delta(r_0 - r)\delta(\Omega_0 - \Omega) \tag{3.133}$$

and Eq. (3.110) is proportional to the field at the input face of the GRIN lens.

For Fourier transform planes $\tilde{z}_p$ such that $H_f(\tilde{z}_p) = 0$ or $u(\tilde{z}_p) = 0$, Eq. (3.111) reduces to

$$K(r_0, r, \Omega_0, \Omega; \tilde{z}_p) = \frac{kn_0}{i2\pi H_a(\tilde{z}_p)} \exp\{ikn_0 \tilde{z}_p\} \cdot \exp\left\{-i\frac{kn_0}{H_a(\tilde{z}_p)} rr_0 \cos(\Omega_0 - \Omega)\right\} \tag{3.134}$$

and Eq. (3.110) becomes

$$\psi(r; \tilde{z}_p) = \frac{kn_0 a^2}{i2H_a(\tilde{z}_p)} \exp\{ikn_0 \tilde{z}_p\} \frac{2J_1[v(\tilde{z}_p)]}{v(\tilde{z}_p)} \tag{3.135}$$

where J_1 is the first-order Bessel function of the first kind.

The irradiance on the Fourier planes is written as

$$I(r; \tilde{z}_p) = I_0(\tilde{z}_p)\left(\frac{2J_1[v(\tilde{z}_p)]}{v(\tilde{z}_p)}\right)^2 \tag{3.136}$$

where

$$I_0(\tilde{z}_p) = I(0; \tilde{z}_p) = \frac{(kn_0)^2 a^4}{4H_a^2(\tilde{z}_p)} \tag{3.137}$$

is the irradiance for on-axis points.

From Eq. (3.136) it follows that we obtain the Airy formula for Fraunhofer diffraction at a circular aperture, as was to be expected. Thus the first minimum (irradiance zero) in the Fourier planes is given by

$$r = 0.61 \frac{\lambda H_a(\tilde{z}_p)}{n_0 a} \tag{3.138}$$

For a quarter-pitch selfoc lens ($g_0 d = \frac{\pi}{2}$), Eq. (3.138) becomes

$$r = \frac{0.61\lambda}{n_0 g_0 a} \tag{3.139}$$

where Eq. (1.96) has been used.

In this case, a focused spot is formed on the output face of the lens, and the focused spot size is limited by diffraction. The spot radius for a 0.25–pitch selfoc lens with $n_0 = 1.591$, $g_0 = 0.199\,\text{mm}^{-1}$ and $a = 1.5\,\text{mm}$ is 1.67 μm with illumination wavelength $\lambda = 1.3\,\mu\text{m}$.

Furthermore, for plane illumination, the irradiance on–axis of the GRIN lens and the positions of the zeros of irradiance, given by Eqs. (3.127) and (3.130), reduce to

$$I(0;z) = \frac{4}{H_f^2(z)} \sin^2\left[\frac{u(z)}{4}\right] \tag{3.140}$$

$$\cot\left[\int_0^{z_q^{oa}} g(z')dz'\right] = \frac{2q\lambda}{n_0 g_0 a^2} \tag{3.141}$$

For optical wavelengths, taking into account the slow variation condition of g(z), Eq. (3.141) can be approximated by

$$\int_0^{z_q^{oa}} g(z')dz' \cong \frac{\pi}{2} - \frac{2q\lambda}{n_0 g_0 a^2} \tag{3.142}$$

If z_m represents an image plane location given by Eq. (2.39), the first zero from this plane is given by

$$\int_{z_m}^{z_q^{oa}} g(z')dz' = -\frac{2\lambda}{n_0 g_0 a^2} \tag{3.143}$$

We assume that both planes are very close; then g(z) may be assumed to have the same value $g(z_m)$ for points on one and the same plane. Hence the first zero of irradiance on the axis is at distance

$$|\Delta z| = \frac{2\lambda}{n_0 a^2 g_0 g(z_m)} \tag{3.144}$$

from the image plane z_m where $\Delta z = z_q^{oa} - z_m$.

In a similar way, if $g(\tilde{z}_p)$ is the value of g(z) in the Fourier plane $\tilde{z}_p$, the first zero of irradiance on the axis is at distance

$$|\Delta \tilde{z}| = \frac{2\lambda}{n_0 a^2 g_0 g(\tilde{z}_p)} \tag{3.145}$$

from the Fourier plane $\tilde{z}_p$, where Eq. (2.38) has been used.

Figure 3.15 shows the irradiance along the z–axis in a selfoc lens illuminated by a uniform plane wave of unit amplitude. The first two focused spots, obtained as $H_f(\tilde{z}_p) = 0$, are represented, and an enlarged region of on-axis irradiance is depicted to illustrate the sinusoidal variation of the irradiance.

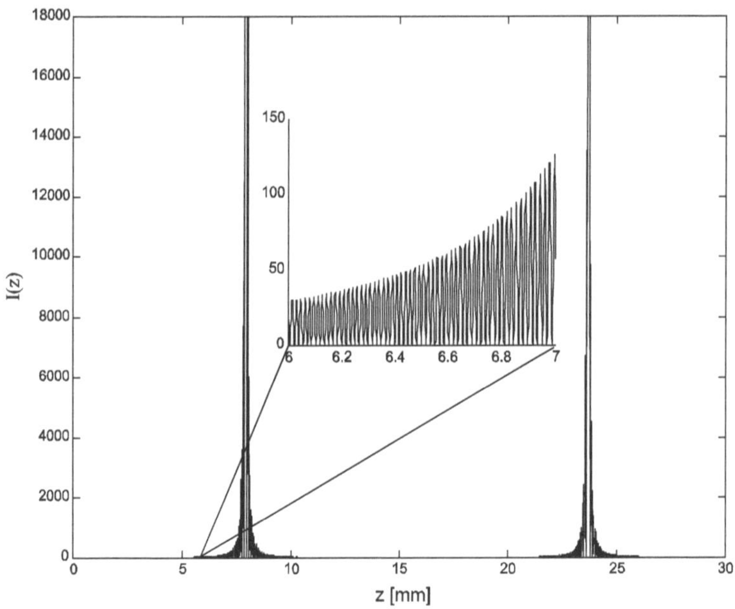

Fig 3.15. On-axis irradiance in a selfoc lens illuminated by a uniform plane wave. Calculations have been made for $n_0 = 1.591$, $g_0 = 0.199\,\text{mm}^{-1}$, $\lambda = 1.3\,\mu\text{m}$ and $a = 1.5\,\text{mm}$.

Finally, note that calculation of the complex amplitude distribution in the geometrically illuminated region and in the geometrical shadow by Lommel functions for a GRIN lens is a similar problem to knowing the three–dimensional light distribution near the focus when estimating the tolerance in the setting of the receiving plane in an image–forming system [3.25].

3.7 Effect of the Aperture on Image and Fourier Transform Formation

We shall now study the effect of the finite extent of the GRIN lens aperture on image and Fourier transform formation. Referring to Fig. 3.5, the complex amplitude distribution at the observation plane is expressed as

$$\psi_0^{dl}(x_1', y_1') = \int_{\Re^2} K_{dl}(x_1, y_1, x_1', y_1'; d_1', d, d_1)\psi_i(x_1, y_1)\, dx_1 dy_1 \qquad (3.146)$$

where K_{dl} is the impulse response of the optical system for diffraction–limited propagation, and ψ_i is the complex amplitude distribution on the input plane, given by Eq. (3.19).

In order to take account of the effect of the finite cross-section of the GRIN lens on image formation, we use the Rayleigh theory for optical systems in which diffraction effects results from the finite size of the output effective aperture [3.26].

As mentioned above, in a GRIN lens, for spherical illumination, the output effective radius is given by

$$a_e'^2 = a_e^2 |F(d)|^2 \qquad (3.147)$$

where a_e and F(d) are given by Eqs. (3.89c) and (3.92), respectively.

The impulse response of the system at the output face of the lens, taking into account Eq. (3.47), is written as

$$K(x_1, y_1, x, y; d_1, d) = \int_{\Re^2} K^{FP}(x_1, y_1, x_0, y_0; d_1) K^{GL}(x_0, y_0, x, y; d)\, dx_0 dy_0 \qquad (3.148)$$

where

$$K^{FP}(x_1, y_1, x_0, y_0; d_1) = \frac{1}{i d_1 \lambda} \exp\left\{ i \frac{k}{2d_1} \left[(x_0 - x_1)^2 + (y_0 - y_1)^2 \right] \right\} \qquad (3.149)$$

is the impulse response over a distance d_1 in free space, and K^{GL} is the impulse response of the lens given by Eq. (2.20).

Inserting Eqs. (2.20) and (3.149) into Eq. (3.148) and integrating, we have

$$K(x_1, y_1, x, y; d_1, d) = \frac{1}{id_1 F(d)\lambda} \exp\left\{i \frac{kH_f(d)}{2d_1 F(d)}(x_1^2 + y_1^2)\right\}$$
$$\exp\left\{i \frac{kn_0 \dot{F}(d)}{2F(d)}(x^2 + y^2)\right\} \exp\left\{-i \frac{k}{d_1 F(d)}(xx_1 + yy_1)\right\} \qquad (3.150)$$

On the output face of the GRIN lens, the complex amplitude distribution is limited by the effective aperture size. Then the impulse response of the optical system at the observation plane, including the effective finite size of the lens, can be written as

$$K_{dl}(x_1, y_1, x_1', y_1'; d_1', d, d_1) = \int_{\Re^2} P(x, y) K(x_1, y_1, x, y; d_1, d) \cdot K^{FP}(x, y, x_1', y_1'; d_1') dxdy \qquad (3.151)$$

where

$$K^{FP}(x_1, y_1, x_1', y_1'; d_1') = \frac{1}{id_1'\lambda} \exp\left\{i \frac{k}{2d_1'}\left[(x_1' - x)^2 + (y_1' - y)^2\right]\right\} \qquad (3.152)$$

is the impulse response over a distance d_1' in free–space, and

$$P(x, y) = \begin{cases} 1 & \text{if} \quad x^2 + y^2 \le a_e'^2(d) \\ 0 & \text{otherwise} \end{cases} \qquad (3.153)$$

is the exit pupil function of the GRIN lens. The finite extent of the GRIN lens has been accounted for by associating a pupil function with the lens, as discussed in Sect. 3.4.

At the image plane given by condition (3.27), the impulse response of the optical system reduces to

$$K_{dl}(x_1, y_1, x_1', y_1'; d_1', d, d_1)$$
$$= \frac{m_t}{d_1'^2 \lambda^2} \exp\left\{-i \frac{km_t H_f(d)}{2d_1'}\left(x_1^2 + y_1^2\right)\right\} \exp\left\{i \frac{k}{2d_1'}\left(x_1'^2 + y_1'^2\right)\right\}$$
$$\cdot \int_{\Re^2} P(x, y) \exp\left\{-i \frac{k}{d_1'}\left[(x_1' - m_t x_1)x + (y_1' - m_t y_1)y\right]\right\} dxdy \qquad (3.154)$$

where m_t is the transverse magnification of the system given by Eq. (3.28).

Hence K_{dl} is proportional to the Fraunhofer diffraction pattern produced by the effective output radius of the lens, centered at the point (mx_1, my_1) of the image plane.

For the sake of simplicity, if we introduce the following change of variables

$$\bar{x} = \frac{x}{\lambda d_1'} \qquad (3.155a)$$

$$\bar{y} = \frac{y}{\lambda d_1'} \tag{3.155b}$$

$$\bar{x}_1 = m_t x_1 \tag{3.155c}$$

$$\bar{y}_1 = m_t y_1 \tag{3.155d}$$

the impulse response becomes

$$K_{dl}\left(\bar{x}_1, \bar{y}_1, x_1', y_1'; d_1', d, d_1\right)$$
$$= m_t \exp\left\{-i\frac{kH_f(d)}{2m_t d_1'}\left(\bar{x}_1^2 + \bar{y}_1^2\right)\right\}\exp\left\{i\frac{k}{2d_1'}\left(x_1'^2 + y_1'^2\right)\right\}$$
$$\cdot \int_{\Re^2} P(\lambda d_1'\bar{x}, \lambda d_1'\bar{y}) \exp\{-i2\pi[(x_1' - \bar{x}_1)\bar{x} + (y_1' - \bar{y}_1)\bar{y}]\}\, d\bar{x}d\bar{y} \tag{3.156}$$

Here

$$P\left(\lambda d_1'\bar{x}, \lambda d_1'\bar{y}\right) = \begin{cases} 1 & \text{if} \quad \bar{x}^2 + \bar{y}^2 \le \dfrac{a_e'^2(d)}{\lambda^2 d_1'^2} = \bar{a}_e'^2(d) \\ 0 & \text{otherwise} \end{cases} \tag{3.157}$$

If we now redefine the impulse response of the optical system for diffraction-limited propagation as the Fourier transform of the reduced effective output radius $\bar{a}_e'$, that is

$$\bar{K}_{dl}(\bar{x}_1, \bar{y}_1, x_1', y_1'; d_1', d, d_1) = \int_{\Re^2} P(\lambda d_1'\bar{x}, \lambda d_1'\bar{y})$$
$$\exp[-i2\pi[(x_1' - \bar{x}_1)\bar{x} + (y_1' - \bar{y}_1)\bar{y}]]\, d\bar{x}d\bar{y} \tag{3.158}$$

the integral transform (3.146) can then be expressed as

$$\psi_0^{dl}(x_1', y_1') = \frac{1}{d_0 m_t}\exp\left\{i\frac{k}{2d_1'}\left(x_1'^2 + y_1'^2\right)\right\}\int_{\Re^2} \bar{K}_{dl}\left(\bar{x}_1, \bar{y}_1, x_1', y_1'; d_1', d, d_1\right)$$
$$\cdot \exp\left\{-i\frac{kH_f(d)}{2m_t d_1'}\left(\bar{x}_1^2 + \bar{y}_1^2\right)\right\}\exp\left\{i\frac{k}{2d_0 m_t^2}\left(\bar{x}_1^2 + \bar{y}_1^2\right)\right\} \tag{3.159}$$
$$f_s\left(\frac{\bar{x}_1}{m_t}, \frac{\bar{y}_1}{m_t}\right) d\bar{x}_1 d\bar{y}_1$$

where Eq. (3.19) has been used.

Likewise, Eq. (3.159) is rewritten symbolically as

$$\psi_0^{dl}(x_1', y_1') = \left\{ \frac{1}{d_0 m_t} \exp\left\{ i \frac{k}{2d_1'} (x_1'^2 + y_1'^2) \right\} \exp\left\{ i \frac{k}{m_t} \left[\frac{1}{d_0 m_t} - \frac{H_f(d)}{d_1'} \right] (x_1'^2 + y_1'^2) \right\} \cdot \right.$$

$$\left. \cdot f_s\left(\frac{x_1'}{m_t}, \frac{y_1'}{m_t} \right) \right\} \otimes \overline{K}_{dl}(x_1', y_1') \tag{3.160}$$

where a $\otimes$ symbol between any two functions denotes the convolution operation.
Note that the first function in the convolution coincides with Eq. (3.30) since

$$\frac{1}{d_1'}[m_t - H_f(d)] = n_0 \dot{H}_f(d) \tag{3.161}$$

where Eq. (3.28) has been used.
Thus the complex amplitude distribution at the image plane is given by

$$\psi_0^{dl}(x_1', y_1') = \psi_0(x_1', y_1') \otimes \overline{K}_{dl}(x_1', y_1') \tag{3.162}$$

We may regard the image for diffraction-limited propagation as a two-dimensional convolution of the image for diffraction-free propagation, predicted by geometrical optics, with the impulse response of the system that is determined by the effective output radius of the GRIN lens. The impulse response is simply the Airy pattern.

We deal now with Fourier transform formation by a GRIN lens. As in image formation, the Rayleigh theory will be used. The complex amplitude distribution at the GRIN lens exit face is given by

$$\psi_0(x, y) = \int_{\Re^2} K(x_1, y_1, x, y; d_1, d) \psi_0(x_1, y_1) \, dx_1 dy_1 \tag{3.163}$$

Inserting Eqs. (3.19) and (3.150) into Eq. (3.163), we have

$$\begin{aligned} \psi_0(x, y) = {} & \frac{1}{i d_1 d_0 \lambda F(d)} \exp\left\{ i \frac{k n_0 \dot{F}(d)}{2F(d)} (x^2 + y^2) \right\} \\ & \cdot \int_{\Re^2} f_s(x_1, y_1) \exp\left\{ i \frac{k}{2} \left[\frac{1}{d_0} + \frac{H_f(d)}{d_1 F(d)} \right] (x_1^2 + y_1^2) \right\} \\ & \cdot \exp\left\{ -i \frac{k}{d_1 F(d)} (x x_1 + y y_1) \right\} dx_1 dy_1 \end{aligned} \tag{3.164}$$

Then, the complex amplitude distribution at the observation plane, a distance d_1' behind the GRIN lens (Fig. 3.5), can be written as

$$\psi_0^{dl}(x_1', y_1') = \int_{\Re^2} P(x, y) \psi_0(x, y) K^{FP}(x, y, x_1', y_1'; d_1') \, dx dy \tag{3.165}$$

where the effective finite size of the GRIN lens is included.

Substitution of Eqs. (3.152) and (3.164) into Eq. (3.165) provides

$$\psi_0^{dl}(x_1', y_1') = -\frac{1}{d_0 d_1 d_1' \lambda^2 F(d)} \exp\left\{ i \frac{k}{2d_1'} \left(x_1'^2 + y_1'^2 \right) \right\} \cdot \int_{\Re^2} f_s(x_1, y_1) \exp\left\{ i \frac{k}{2} \left[\frac{1}{d_0} + \frac{H_f(d)}{d_1 F(d)} \right] \left(x_1^2 + y_1^2 \right) \right\} \cdot H(x_1', y_1'; x_1, y_1) dx_1 dy_1 \tag{3.166}$$

where

$$H(x_1', y_1'; x_1, y_1) = \int_{\Re^2} P(x, y) \exp\left\{ i \frac{k}{2} \left[\frac{1}{d_1'} + \frac{n_0 \dot{F}(d)}{F(d)} \right] \left(x^2 + y^2 \right) \right\} \cdot \exp\left\{ -i \frac{k}{d_1'} \left[\left(x_1' + \frac{d_1'}{d_1 F(d)} x_1 \right) x + \left(y_1' + \frac{d_1'}{d_1 F(d)} y_1 \right) y \right] \right\} dx dy \tag{3.167}$$

The above equation can be written in polar coordinates as

$$H(x_1', y_1'; x_1, y_1) = 2\pi \int_0^{a_e'(d)} \exp\left\{ i \frac{k}{2} \left[\frac{1}{d_1'} + \frac{n_0 \dot{F}(d)}{F(d)} \right] r^2 \right\} J_0\left(\frac{kr'r}{d_1'} \right) r dr \tag{3.168}$$

where

$$r^2 = x^2 + y^2 \tag{3.169}$$

$$r'^2 = \left(x_1'^2 + \frac{d_1'}{d_1 F(d)} x_1 \right)^2 + \left(y_1'^2 + \frac{d_1'}{d_1 F(d)} y_1 \right)^2 \tag{3.170}$$

and Eq. (3.116) has been used.

Equation (3.168) may be evaluated in terms of the Lommel functions as mentioned in Sect. 3.5. Here dimensionless variables are given by

$$u = \frac{k[F(d) + n_0 d_1' \dot{F}(d)] a_e'^2(d)}{d_1' F(d)} \tag{3.171}$$

$$v = \frac{kr' a_e'(d)}{d_1'} \tag{3.172}$$

$$w = \frac{r}{a_e'(d)} \tag{3.173}$$

and

$$\frac{u}{v} = \frac{F(d) + n_0 d_1' \dot{F}(d)}{r' F(d)} a_e'(d) = \frac{[F(d) + n_0 d_1' \dot{F}(d)] a_e}{r'} \tag{3.174}$$

where Eq. (3.147) has been used.

The numerator of Eq. (3.174) represents the position, at the observation plane, of the marginal ray emerging from the point a_e' on the output face of the GRIN lens, and then $|u/v|$ defines the geometrically illuminated and geometrical shadow regions on the observation plane.

Equation (3.168) in terms of the Lommel functions becomes

$$H(x_1', y_1'; x_1, y_1) = \pi a_e'^2 [C(u, v) + iS(u, v)] \tag{3.175}$$

On the other hand, the general condition to obtain the Fourier transform for free propagation given by Eq. (3.36) can be expressed as

$$d_0 = -\frac{F(d) + n_0 d_1' \dot{F}(d)}{H_f(d) + n_0 d_1' \dot{H}_f(d)} d_1 \tag{3.176}$$

Inserting Eqs. (3.176) and (3.119) into Eq. (3.175), we obtain for the resulting complex amplitude distribution at the observation plane that

$$\psi_0^{dl}(x_1', y_1') = \psi_0(x_1', y_1') + \tilde{\psi}_0(x_1', y_1') \quad \text{if} \quad \left|\frac{u}{v}\right| \rangle 1 \tag{3.177}$$

where ψ_0 is the Fourier transform of the transparency f_s for free propagation given by Eq. (3.37), and

$$\tilde{\psi}_0(x_1', y_1') = \frac{k}{2\pi d_0 d_1 [F(d) + n_0 d_1' \dot{F}(d)]} \cdot$$

$$\exp\left\{i \frac{k}{2d_1'}(x_1'^2 + y_1'^2)\right\} \exp\left\{i \frac{k}{2}\left[\frac{1}{d_1'} + \frac{n_0 \dot{F}(d)}{F(d)}\right] a_e'^2(d)\right\} \cdot$$
$$\int_{\Re^2} f_s(x_1, y_1) \exp\left\{i \frac{k d_1'}{2 d_1^2}(x_1^2 + y_1^2)\right\} [iV_0(u, v) + V_1(u, v)] dx_1 dy_1 \tag{3.178}$$

Thus for the geometrically illuminated region, the resulting field may be regarded as the sum of the Fourier transform of the transparency and another term arising from the effects of diffraction due to the effective output radius.

In the same way, if $|u/v| = 1$ we have

$$\psi_0^{dl}(x_1', y_1') = \frac{ik\left[H_f(d) + n_0 d_1' \dot{H}_f(d)\right]}{4\pi d_1^2 \left[F(d) + n_0 d_1' \dot{F}(d)\right]^2} \exp\left\{i\frac{u}{2}\right\}\left\{\exp[-iu] - J_0(u)\right\} \tag{3.179}$$
$$\exp\left\{i\frac{k}{2d_1'}\left(x_1'^2 + y_1'^2\right)\right\} \int_{\Re^2} f_s(x_1, y_1) \exp\left\{i\frac{kd_1'}{2d_1^2}\left(x_1^2 + y_1^2\right)\right\} dx_1 dy_1$$

For the edge of the geometrical shadow, the resulting complex amplitude distribution can be regarded as being due to diffractional effects of the finite extent of the GRIN lens.

If $\left|\frac{u}{v}\right| < 1$ we get

$$\psi_0^{dl}(x_1', y_1') = \frac{ik\left[H_f(d) + n_0 d_1' \dot{H}_f(d)\right]}{2\pi d_1^2 \left[F(d) + n_0 d_1' \dot{F}(d)\right]^2} \exp\left\{i\frac{u}{2}\right\} \exp\left\{i\frac{k}{2d_1'}\left(x_1'^2 + y_1'^2\right)\right\} \tag{3.180}$$
$$\int_{\Re^2} f_s(x_1, y_1) \exp\left\{i\frac{kd_1'}{2d_1^2}\left(x_1^2 + y_1^2\right)\right\} \left[U_1(u, v) - iU_2(u, v)\right] dx_1 dy_1$$

Hence for the geometrical shadow, the complex amplitude distribution at the observation plane is given by diffraction–limited propagation through the optical system.

Likewise, the size of the geometrically illuminated region on the observation plane can be evaluated from the edge of the geometrical shadow condition

$$|u|^2 = |v|^2 \tag{3.181}$$

Inserting Eqs. (3.171–3.172) and taking into account Eq. (3.170), we obtain

$$\left(x_1' + \frac{d_1'}{d_1 F(d)} x_1\right)^2 + \left(y_1' + \frac{d_1'}{d_1 F(d)} y_1\right)^2 = [F(d) + n_0 d_1' \dot{F}(d)]^2 a_e'^2(d) \tag{3.182}$$

This region will be a circle of radius

$$r' = [F(d) + n_0 d_1' \dot{F}(d)] a_e'(d) \tag{3.183}$$

and centered at the point having coordinates

$$x_{c1}' = -\frac{d_1'}{d_1 F(d)} x_1 \tag{3.184a}$$

$$y_{c1}' = -\frac{d_1'}{d_1 F(d)} y_1 \tag{3.184b}$$

We can also obtain the size of the effective object. From Eq. (3.182) it follows that the effective portion of f_s that contributes to the resulting complex amplitude distribution is a circle of radius

$$r_s = \frac{d_1[F(d) + n_0 d_1' \dot{F}(d)] a_e'(d) F(d)}{d_1'} \quad (3.185)$$

centered at coordinates

$$x_{c1} = -\frac{d_1 F(d)}{d_1'} x_1' \quad (3.186a)$$

$$y_{c1} = -\frac{d_1 F(d)}{d_1'} y_1' \quad (3.186b)$$

Thus, the part of the object that contributes to the Fourier transform depends on the coordinates (x_1', y_1') being considered in the observation plane [3.6]. The limitation of the spatial extent of the object by the finite lens aperture causes vignetting as mentioned in Sect. 3.5.

4 GRIN Lenses for Gaussian Illumination

4.1 Introduction

In Chap. 3, we discussed behavior of GRIN lenses illuminated by a uniform monochromatic wave. But, it is well–known that the output modes of most lasers can be simply described by Hermite–Gaussian or Laguerre–Gaussian functions. Therefore, for applications involving the propagation of laser beams, it is important to understand the effect of GRIN lenses on such Gaussian beams since these lenses are used, for instance, in optical fiber communications and optoelectronic systems for manipulating and processing optical signals.

Chapter 4 summarizes some useful properties of Gaussian beams and studies the laws of transformation of these beams through and by GRIN lenses. Chapter 4 also discusses other related topics for non–uniform monochromatic waves described by Gaussian beams.

4.2 Propagation of Gaussian Beams in a GRIN Lens

We consider a GRIN lens with rotational symmetry around the z–axis whose refractive index is given by Eq. (3.1). We shall now study light propagation in a GRIN lens when it is illuminated by a non-uniform monochromatic wave of wavelength λ, described by a Gaussian beam. That is, we are concerned with waves whose irradiance is maximum on the axis and decreases as a Gaussian function with distance from the axis. This is why the waves are called Gaussian beams. These beams are solutions of the scalar parabolic wave equation, and they are represented by the Hermite–Gauss or Laguerre–Gauss modes of lowest–order [4.1].

When a GRIN lens is illuminated, for instance, by a spherical Gaussian beam (Fig. 4.1), the complex amplitude at the input face can be written in the paraxial region by [4.2–4.4]

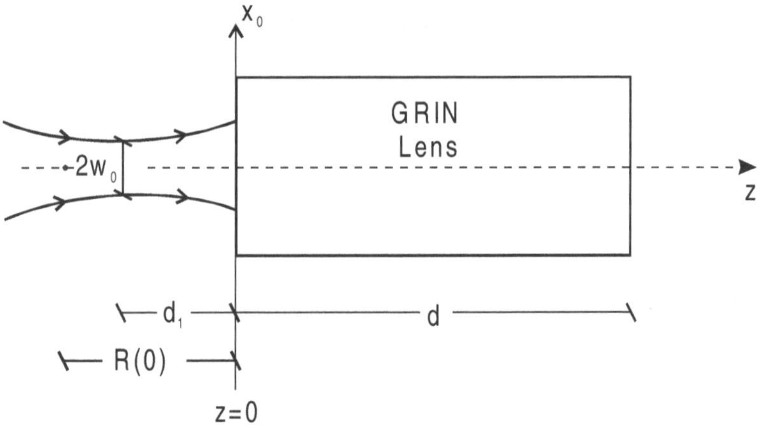

Fig. 4.1. Geometry for the evaluation of the complex amplitude distribution inside a GRIN lens illuminated by a Gaussian beam, where R(0), w_0, and d_1 are the curvature radius, the beam waist, and the distance from the waist plane to the input face of the GRIN lens, respectively.

$$\psi(x_0, y_0; 0) = \frac{w_0}{w(0)} \exp\{i\varphi(0)\} \exp\left\{ i \frac{\pi U(0)}{\lambda} \left(x_0^2 + y_0^2\right) \right\} \tag{4.1}$$

where the beam parameters in free space at a distance d_1 from the waist plane of diameter or spot size $2w_0$ are given by the complex wavefront curvature

$$U(0) = \frac{1}{R(0)} + i\frac{\lambda}{\pi w^2(0)} \tag{4.2}$$

and the on–axis phase

$$\varphi(0) = -\tan^{-1}\left(\frac{\lambda d_1}{\pi w_0^2}\right) \tag{4.3}$$

where R(0) and w(0) are the radius of curvature and the beam half–width or beam radius at $z = 0$. The beam half–width is the distance from the axis at which the irradiance decays to e^{-2} of its maximum value [4.5]. Complex wavefront curvature reduces to real wavefront curvature in the geometrical optics limit, $k \to \infty$.

Likewise, the relationship between the beam radius at $z = 0$ and the waist radius is governed by

$$w(0) = w_0 \left[1 + \left(\frac{d_1}{z_0}\right)^2\right]^{1/2} \tag{4.4}$$

where

$$z_0 = \frac{\pi w_0^2}{\lambda} \tag{4.5}$$

is the Rayleigh range characterizing the Lorentzian profile of irradiance along the axis.

The complex amplitude distribution in the GRIN lens is given by the integral transform from Eq. (2.1)

$$\psi(x,y;z) = \int_{\Re^2} \psi(x_0,y_0;0)K(x_0,y_0,x,y;z)\,dx_0 dy_0 \tag{4.6}$$

where K is the kernel function.

Substituting Eq. (4.1) into Eq. (4.6) and integrating, we have

$$\psi(x,y;z) = \frac{w_0}{w(0)G(z)}\exp\{i\varphi(z)\}\exp\left\{\frac{i\pi n_0 \dot{G}(z)}{\lambda G(z)}(x^2+y^2)\right\} \tag{4.7}$$

where Eq. (2.20) has been used.

Equation (4.7) is the central result of the present analysis, and it represents a spherical Gaussian beam of complex curvature.

$$U(z) = n_0\frac{d}{dz}[\ln G(z)] = n_0\dot{G}(z)G^{-1}(z) \tag{4.8}$$

and the on–axis phase

$$\varphi(z) = \varphi(0) + kn_0 z \tag{4.9}$$

where

$$\overset{(\cdot)}{G}(z) = \overset{(\cdot)}{H_f}(z) + \frac{U(0)}{n_0}\overset{(\cdot)}{H_a}(z) \tag{4.10}$$

The complex curvature may also be expressed as [4.6]

$$U(z) = \frac{1}{R(z)} + i\frac{\lambda}{\pi w^2(z)} \tag{4.11}$$

where R(z) and w(z) are the radius of curvature and the beam half–width of the Gaussian beam in the GRIN lens at z.

When Eq. (4.8) is compared with Eq. (4.11) and Eqs. (4.2) and (4.10) are taken into account, it follows that R(z) and w(z) are given by

$$\begin{aligned}\frac{1}{R(z)} = \frac{1}{|G(z)|^2}\Bigg\{&\frac{1}{R(0)}\left[H_a(z)\dot{H}_f(z)+\dot{H}_a(z)H_f(z)\right]\\ &+\left[\frac{1}{z_R^2}+\frac{1}{n_0^2R^2(0)}\right]n_0H_a(z)\dot{H}_a(z)+n_0H_f(z)\dot{H}_f(z)\Bigg\}\end{aligned} \tag{4.12}$$

$$w^2(z) = w^2(0)|G(z)|^2 \tag{4.13}$$

where z_R is the Rayleigh range along the z–axis of the GRIN lens, expressed as

$$z_R = \frac{\pi n_0 w^2(0)}{\lambda} \tag{4.14}$$

and $|G|$ is the modulus of G(z); that is

$$|G(z)|^2 = \left[\frac{H_a(z)}{n_0 R(0)} + H_f(z)\right]^2 + \frac{H_a^2(z)}{z_R^2} = G_r^2(z) + G_i^2(z) \tag{4.15}$$

which relates the beam radius at z > 0 to the beam radius at the input face of the GRIN lens, where G_r and G_i are the real and imaginary parts of G, respectively.

Likewise, from Eqs. (4.8) and (4.13) it follows that the complex amplitude distribution of the Gaussian beam in the GRIN lens can be written as

$$\psi(x,y;z) = \frac{w_0}{w(z)} \exp\{i\varphi(z)\} \exp\left\{-i\tan^{-1}\left[G_i(z)/G_r(z)\right]\right\} \exp\left\{i\frac{\pi U(z)}{\lambda}(x^2+y^2)\right\} \tag{4.16}$$

Equations (4.7) or (4.16) include plane Gaussian illumination as a particular case. For this illumination, the waist w_0 is located at the input face of the GRIN lens and $R(0) \to \infty$. Under these conditions $\overset{(\cdot)}{G}(z)$ becomes

$$\overset{(\cdot)}{G}_p(z) = \overset{(\cdot)}{H}_f(z) + i\frac{\overset{(\cdot)}{H}_a(z)}{z_{pR}} \tag{4.17}$$

where

$$z_{pR} = \frac{\pi n_0 w_0^2}{\lambda} \tag{4.18}$$

and the radius of curvature, beam profile, and complex amplitude distribution for plane illumination are now given by

$$\frac{1}{R_p(z)} = \frac{n_0}{|G_p(z)|^2}\left[\frac{H_a(z)\dot{H}_a(z)}{z_{pR}^2} + H_f(z)\dot{H}_f(z)\right] \tag{4.19}$$

$$w_p^2(z) = w_0^2 |G_p(z)|^2 \tag{4.20}$$

$$\psi_P(x,y;z) = \frac{1}{|G_p(z)|} \exp\{i\varphi_p(z)\} \exp\left\{-i\tan^{-1}\left[\frac{H_a(z)}{z_{pR} H_f(z)}\right]\right\} \exp\left\{i\frac{\pi U_p(z)}{\lambda}(x^2+y^2)\right\} \tag{4.21}$$

where

$$\varphi_p(z) = kn_0 z - \tan^{-1}\left(\frac{n_0 d_1}{z_{pR}}\right) \tag{4.22}$$

$$U_p(z) = n_0 \dot{G}_p(z) G_p^{-1}(z) \tag{4.23}$$

On the other hand, as mentioned in Chap. 2, there is a connection between the kernel function and the ray–transfer matrix that leads to useful relations between the geometrical ray optics and the wave optics. In this way, the passage of a Gaussian beam through a GRIN lens is described by the ABCD law [4.7]

$$\begin{pmatrix} q(z) \\ n_0\dot{q}(z) \end{pmatrix} = \begin{pmatrix} A' & B' \\ C' & D' \end{pmatrix} \begin{pmatrix} q(0) \\ n_0\dot{q}(0) \end{pmatrix} \tag{4.24}$$

where the elements of the ray–transfer matrix are given by Eqs. (1.91b).

In this case, the ABCD law relate complex rays whose positions and optical direction cosines are denoted by $q(0)$, $n_0\dot{q}(0)$ and q, $n_0\dot{q}$ at the planes $z = 0$ and $z > 0$, respectively. The complex parameters q and $\dot{q}$, resulting from real rays by means of analytic continuation, describe Gaussian beam propagation and play a role similar to the one played by the position and slope of the real geometrical rays used in uniform illumination.

The complex beam parameter $q(z)$ obeys the paraxial ray equation, and the real and imaginary parts q_r and q_i, respectively, are also two solutions of the equation [4.1, 4.2]

$$\ddot{q}_{\binom{r}{i}} + g^2(z) q_{\binom{r}{i}} = 0 \tag{4.25}$$

satisfying the z–invariant condition

$$q_r\dot{q}_i - \dot{q}_r q_i = \frac{2}{kn_0} \tag{4.26}$$

or equivalently

$$q^*\dot{q} - q\dot{q}^* = 2i(q_r\dot{q}_i - \dot{q}_r q_i) = \frac{4i}{kn_0} \tag{4.27}$$

In order to find the relations between the geometrical ray optics and the wave optics for Gaussian beam propagation in a GRIN lens, Eq. (4.24) is written as

$$q(z) = q(0)[A' + n_0\dot{q}(0)q^{-1}(0)B'] \tag{4.28a}$$

$$\dot{q}(z) = q(0)\left[\frac{C'}{n_0} + \dot{q}(0)q^{-1}(0)D'\right] \tag{4.28b}$$

At the input face of the GRIN lens, position and slope of the complex rays are given by

$$q(0) = w_0 + i\frac{\lambda d_1}{\pi w_0} \tag{4.29a}$$

$$\dot{q}(0) = \frac{i\lambda}{\pi w_0 n_0} \tag{4.29b}$$

From Eqs. (4.29) and after straightforward calculation we obtain

$$n_0\dot{q}(0)q^{-1}(0) = U(0) \tag{4.30}$$

where the following relationship

$$R(0) = d_1\left[1+\left(\frac{\pi w_0^2}{\lambda d_1}\right)^2\right] \tag{4.31}$$

between R(0) and d_1 has been used, and U(0) is given by Eq. (4.2).

Substitution of Eq. (4.30) into Eqs. (4.28) provides

$$q(z) = q(0)G(z) \tag{4.32a}$$

$$\dot{q}(z) = q(0)\dot{G}(z) \tag{4.32b}$$

where $\overset{(\cdot)}{G}(z)$ coincides with Eq. (4.10). Then, from Eqs. (4.32) the complex curvature of the Gaussian beam in the GRIN lens can be now expressed in terms of q(z) and $\dot{q}(z)$ as

$$U(z) = n_0\dot{q}(z)q^{-1}(z) \tag{4.33}$$

that is, U(z) in terms of the ray matrix elements is written as [4.8–4.14].

$$U(z) = \frac{C' + U(0)D'}{A' + U(0)B'} \tag{4.34}$$

Comparing Eq. (4.33) with Eq. (4.11), the beam half–width and the radius of curvature of the Gaussian beam in terms of q and $\dot{q}$ are given by

$$w^2(z) = \frac{\lambda|q(z)|^2}{n_0\pi[q_r(z)\dot{q}_i(z) - \dot{q}_r(z)q_i(z)]} = |q(z)|^2 \tag{4.35}$$

$$\frac{1}{R(z)} = \frac{n_0[q_r(z)\dot{q}_r(z) + q_i(z)\dot{q}_i(z)]}{|q(z)|^2} = \frac{n_0}{|q(z)|}\frac{d|q(z)|}{dz} = n_0\frac{d}{dz}\ln|q(z)| \tag{4.36}$$

where Eq. (4.26) has been used.

From Eqs. (4.35-38) it follows that the modulus of the complex ray is equal to the beam half-width and that R(z) and w(z) are related by

$$\frac{1}{R(z)} = \frac{n_0}{2w^2(z)}\frac{dw^2(z)}{dz} \tag{4.37}$$

Finally, the on-axis phase of Eq. (4.16) in terms of the complex ray may be expressed as [4.2]

$$kn_0z - \frac{1}{2}\text{phase}\left[\frac{q(z)}{q(0)}\right] = kn_0z + \frac{1}{2}\tan^{-1}\left[\frac{q_i(0)}{q_r(0)}\right] - \frac{1}{2}\tan^{-1}\left[\frac{q_i(z)}{q_r(z)}\right] \tag{4.38}$$

$$= kn_0z + \frac{1}{2}\tan^{-1}\left\{\frac{1}{q_r(0)q_r(z)+q_i(0)q_i(z)}\left[\frac{q_i(0)}{q_r(0)} - \frac{q_i(z)}{q_r(z)}\right]\right\}$$

In short, the wave optics parameters (complex wavefront curvature and on–axis phase) defining the field of a Gaussian beam can be related, in a simple way, to the complex ray q.

Equation (4.13) indicates the evolution of the half-width; we are now interested in those lengths where the half–width is an extremum (beam waist). Then, the condition of the plane Gaussian beam inside the GRIN lens can be obtained by evaluation of the extremum values of the beam half-width or by the vanishing of the wavefront curvature. From Eq. (4.13) or (4.37), this condition provides

$$\frac{dw^2(z)}{dz} = \frac{d|G(z)|^2}{dz} = 0 \tag{4.39}$$

that is,

$$\frac{1}{R(0)}[H_a(z)\dot{H}_f(z) + \dot{H}_a(z)H_f(z)] + \left[\frac{1}{z_R^2} + \frac{1}{n_0^2R^2(0)}\right]n_0H_a(z)\dot{H}_a(z) + n_0H_f(z)\dot{H}_f(z) = 0 \tag{4.40}$$

where Eqs. (4.12) or (4.15) have been used.

Equation (4.40) has two oscillatory solutions z^+ and z^-, in which the beam half–width can be a maximum or a minimum. The axial localizations of these positions can be obtained if we take into account that position and slope of the axial and field rays, given by Eqs. (1.92) and (1.95), are written as

$$H_a(z) = -\frac{\dot{H}_f(z)}{g_0g(z)} = \frac{u(z)}{\{g_0g(z)[1+u^2(z)]\}^{1/2}} \tag{4.41}$$

$$H_f(z) = \frac{g_0}{g(z)}\dot{H}_a(z) = \left\{\frac{g_0}{g(z)[1+u^2(z)]}\right\}^{1/2} \tag{4.42}$$

where

$$u(z) = \tan\left[\int_0^z g(z')\,dz'\right] \tag{4.43}$$

Substituting Eqs. (4.41, 4.42) into Eq. (4.40), we have the following second-order equation

$$au^2(z) + bu(z) + c = 0 \tag{4.44}$$

with

$$a = -c = \frac{g_0}{R(0)} \tag{4.45a}$$

$$b = -n_0\left[\frac{1}{z_R^2} + \frac{1}{n_0^2 R^2(0)} - g_0^2\right] \tag{4.45b}$$

Taking into account Eq. (4.14), solution of Eq. (4.44) is given by

$$u(z^\pm) = \frac{n_0}{2g_0}\left\{\frac{1}{n_0^2 R(0)} + g_0^2 R(0)\left[\frac{w_{fm}^4(0)}{w^4(0)} - 1\right]\right\}$$
$$\pm\left\{\left(\frac{n_0}{2g_0}\right)^2\left[\frac{1}{n_0^2 R(0)} + g_0^2 R(0)\left[\frac{w_{fm}^4(0)}{w^4(0)} - 1\right]\right]^2 + 1\right\}^{1/2} \tag{4.46}$$

where

$$w_{fm}(0) = \left(\frac{\lambda}{\pi n_0 g_0}\right)^{1/2} \tag{4.47}$$

is the half–width of the fundamental mode at z = 0 in a tapered GRIN medium [4.15], and with the use of plus for z^+ and minus for z^-. Note that when $w_{fm}(0) = w(0)$, it follows from Eq. (4.46) that the axial positions of extremum values of the beam half–width reduces to

$$u_p(z^\pm)\Big|_{w_{fm}(0)=w(0)} = \frac{1}{2n_0 g_0 R(0)} \pm \left[\left(\frac{1}{2n_0 g_0 R(0)}\right)^2 + 1\right]^{1/2} \tag{4.48}$$

On the other hand, $|G(z)|$ given by Eq. (4.15) can be expressed in terms of u(z) as

$$|G(z)|^2 = \frac{g_0}{g(z)[1+u^2(z)]}\left[\left(\frac{u(z)}{n_0 g_0 R(0)} + 1\right)^2 + \frac{u^2(z) w_{fm}^4(0)}{w^4(0)}\right] \tag{4.49}$$

and the second derivative of Eq. (4.49) with respect to z is given by

$$\frac{d^2|G(z)|^2}{dz^2} = \frac{2g(z)}{n_0\left[1+u^2(z)\right]}$$
$$\cdot\left\{-4\frac{u(z)}{R(0)} + \frac{n_0}{g_0}\left[1-u^2(z)\right]\left[\frac{1}{n_0^2 R_0^2} + g_0^2\left(\frac{w_{fm}^4(0)}{w^4(0)} - 1\right)\right]\right\} \tag{4.50}$$

It is easy to prove that when $w_{fm}(0)$ is greater than, smaller than, or equal to w(0) and R(0) > 0 (diverging illumination), we have

$$\left.\frac{d^2|G(z)|^2}{dz^2}\right|_{u(z)=u(z^+)} \langle 0, \quad \text{and} \quad \left.\frac{d^2|G(z)|^2}{dz^2}\right|_{u(z)=u(z^-)} \rangle 0 \tag{4.51}$$

Then, the axial positions in which the beam half-width is a maximum or a minimum are given by $u(z^+)$ and $u(z^-)$, respectively.

Conversely, when $w_{fm}(0)$ is greater than, smaller than, or equal to w(0) and R(0)<0 (converging illumination), we arrive at

$$\left.\frac{d^2|G(z)|^2}{dz^2}\right|_{u(z)=u(z^+)} \rangle 0, \quad \text{and} \quad \left.\frac{d^2|G(z)|^2}{dz^2}\right|_{u(z)=u(z^-)} \langle 0 \tag{4.52}$$

The axial locations in which the beam half–width is a maximum or a minimum are obtained for $u(z^-)$ and $u(z^+)$, respectively.

Likewise, it is also of interest to examine the evolution of the beam half–width for plane Gaussian illumination. In this case, for $w_{fm}\langle w_0$ and $w_{fm}\rangle w_0$, Eqs. (4.50) and (4.46) become

$$\left.\frac{d^2|G(z)|^2}{dz^2}\right|_{R(0)\to\infty} = \frac{2g(z)g_0[1-u^2(z)]}{1+u^2(z)}\left\{\frac{w_{fm}^4(0)}{w_0^4}-1\right\} \tag{4.53}$$

$$\lim_{R(0)\to\infty} u(z^+) \to \begin{cases} \infty & \text{for } w_{fm}(0)\rangle w_0 \\ 0 & \text{for } w_{fm}(0)\langle w_0 \end{cases} \tag{4.54}$$

$$\lim_{R(0)\to\infty} u(z^-) \to \begin{cases} 0 & \text{for } w_{fm}(0)\rangle w_0 \\ -\infty & \text{for } w_{fm}(0)\langle w_0 \end{cases} \tag{4.55}$$

From Eqs. (4.54–4.55) and (4.43) it follows that when $w_{fm}\rangle w_0$, we have

$$\cos\left[\int_0^{z^+} g(z')dz'\right] = 0 \quad \text{or} \quad H_f(z^+) = 0, \quad \text{for} \quad u(z^+) \tag{4.56}$$

$$\sin\left[\int_0^{z^-} g(z')dz'\right] = 0 \quad \text{or} \quad H_a(z^-) = 0, \quad \text{for} \quad u(z^-) \tag{4.57}$$

where Eqs. (1.92) have been used.

For Gaussian plane illumination and $w_{fm}(0)\rangle w_0$, the sign of the second derivative of $|G(z)|$ evaluated at $u(z^\pm)$ is given by

$$\left.\frac{d^2|G(z)|^2}{dz^2}\right|_{u(z)=u(z^+)} \langle 0 \tag{4.58}$$

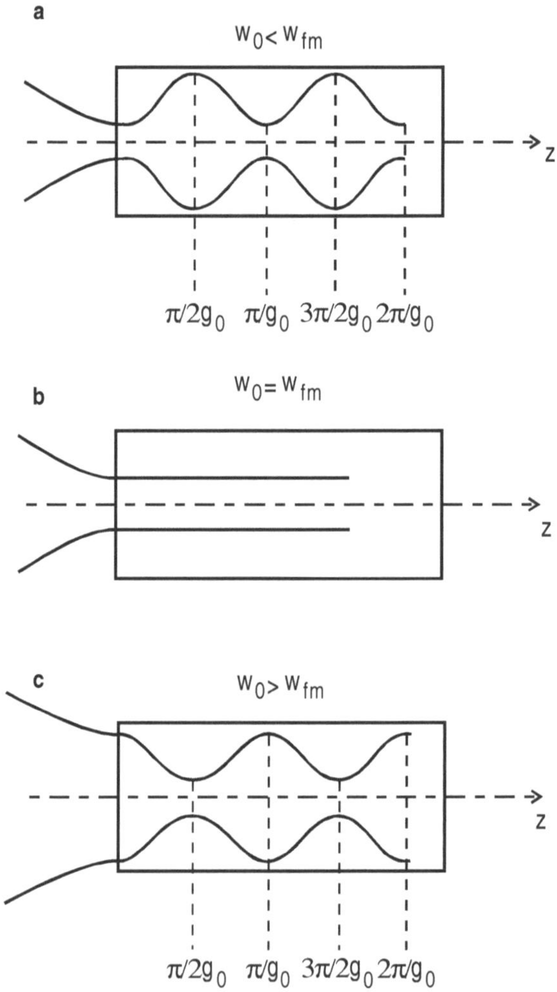

Fig. 4.2. Evolution of the beam half-width inside a selfoc lens illuminated by a plane Gaussian beam for **a** $w_{fm} > w_0$, **b** $w_{fm} = w_0$, and **c** $w_{fm} < w_0$.

$$\left.\frac{d^2|G(z)|^2}{dz^2}\right|_{u(z)=u(z^-)} > 0 \tag{4.59}$$

Consequently, the maximum or minimum values of the beam half–width are achieved at the transforming and imaging conditions respectively, as mentioned in Chap. 2. The reverse behavior occurs for Gaussian plane illumination and $w_{fm}(0) < w_0$.

Finally, note that when $w_{fm}(0) = w_0$, it follows from Eqs. (4.49) and (4.53) that

$$\left.|G(z)|^2\right|_{\substack{R(0)\to\infty \\ w_{fm}(0)=w_0}} = \frac{g_0}{g(z)} \tag{4.60}$$

$$\left.\frac{d^2|G(z)|^2}{dz^2}\right|_{\substack{R(0)\to\infty \\ w_{fm}(0)=w_0}} = 0 \tag{4.61}$$

for all values of z inside the GRIN lens, and the incident plane Gaussian beam propagates as the fundamental mode since

$$\left.w(z)\right|_{\substack{R(0)\to\infty \\ w_{fm}(0)=w_0}} = \left[\frac{g_0}{g(z)}\right]^{1/2} w_{fm}(0) = \left[\frac{\lambda}{\pi n_0 g(z)}\right]^{1/2} = w_{fm}(z) \tag{4.62}$$

where Eqs. (4.20) and (4.47) have been used.

Figure 4.2 depicts the evolution of the beam half-width inside a selfoc lens illuminated by a plane Gaussian beam. When $w_{fm} = w_0$ the beam propagates as the fundamental mode, and there is an adiffractional Gaussian beam inside the selfoc lens.

4.3 GRIN Lens Law: Image and Focal Shifts

We now consider transformation of Gaussian beams by a GRIN lens. Referring to the geometry of Fig. 4.3, and in comparison to the uniform illumination case in calculating the GRIN lens law, we regard the waist of the input beam as the object, and the waist of the output beam as the image. The ratio of the input waist to the output waist is the transverse magnification. Our aim is to determine the condition for which the output beam radius becomes the output waist, that is, for which the beam half-width is a minimum.

We apply the ABCD law to the derivation of the GRIN lens law for Gaussian beams. The ray position q' and the ray slope $\dot{q}'$ at any output plane of the optical system, located at distance d' from the output face of the GRIN lens, are related by the ray position q_0 and slope $\dot{q}_0$ at the waist w_0 of the input Gaussian beam by the matrix equation

$$\begin{pmatrix} q' \\ \dot{q}' \end{pmatrix} = M_s \begin{pmatrix} q_0 \\ \dot{q}_0 \end{pmatrix} \tag{4.63}$$

where

$$q_0 = w_0\,;\; \dot{q}_0 = \frac{i\lambda}{\pi w_0} \tag{4.64}$$

and the elements of the ray–transfer matrix M_s of the optical system are given by Eqs. (3.51) for $n_1 = n_1' = 1$, provided that the GRIN lens is surrounded by free space.

From Eqs. (3.51) and (4.63–4.64) it follows that

$$q' = w_0\left\{H_f(d) + n_0 d'\dot{H}_f(d) + \frac{i}{z_R}\left[H_a(d) + n_0 d'\dot{H}_a(d) + n_0 d_1\left(H_f(d) + n_0 d'\dot{H}_f(d)\right)\right]\right\} \tag{4.65}$$

Then the beam half–width of the Gaussian wavefront at the output plane can be written as

$$w' = |q'| = w_0\left\{\left[H_f(d) + n_0 d'\dot{H}_f(d)\right]^2 + \frac{1}{z_r^2}\left[H_a(d) + n_0 d'\dot{H}_a(d) + n_0 d_1\left(H_f(d) + n_0 d'\dot{H}_f(d)\right)\right]^2\right\}^{1/2} \tag{4.66}$$

The output beam half-width reaches its minimum value at the waist (Fig. 4.3). Thus, the image condition can be expressed as

$$\frac{dw'^2}{dd'} = 0 \tag{4.67}$$

Equation (4.67) indicates that the distance from the output face of the GRIN lens to the waist is given by [4.16].

$$\begin{aligned} d_g' &= -\frac{H_f(d)\dot{H}_f(d)z_R^2 + \left[\dot{H}_a(d) + n_0 d_1\dot{H}_f(d)\right]\left[H_a(d) + n_0 d_1 H_f(d)\right]}{n_0\left[\dot{H}_f^2(d)z_R^2 + \left(\dot{H}_a(d) + n_0 d_1\dot{H}_f(d)\right)^2\right]} \\ &= \frac{1}{2n_0 g^2(d)}\frac{d}{dz}\left\{\ln\left[\dot{H}_f^2(z)z_R^2 + \left(\dot{H}_a(z) + n_0 d_1\dot{H}_f(z)\right)^2\right]\right\}_{z=d} \end{aligned} \tag{4.68}$$

Equation (4.68) represents the final result of the present analysis and can be called the GRIN lens law for Gaussian illumination.

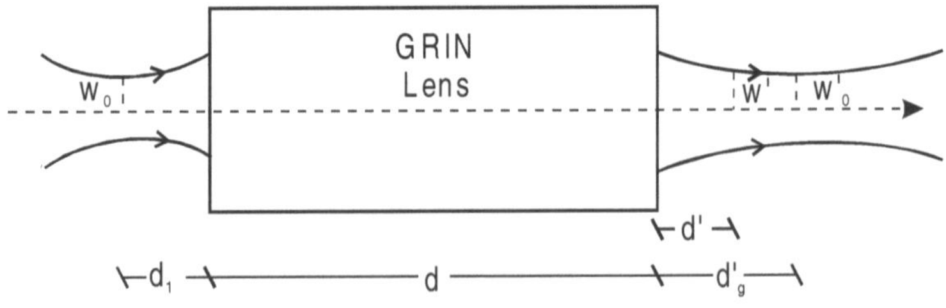

Fig. 4.3. Gaussian beam transformation by a GRIN lens.

Substituting Eq. (4.68) into Eq. (4.66), we obtain the waist of the output beam

$$w_0' = \frac{w_0}{\left[\dot{H}_f^2(z)z_R^2 + (\dot{H}_a(d) + n_0 d_1 \dot{H}_f(d))^2\right]^{1/2}} \tag{4.69}$$

Then, the transverse magnification is given by

$$m_t^g = \frac{w_0'}{w_0} = \left[\dot{H}_f^2(z)z_R^2 + (\dot{H}_a(d) + n_0 d_1 \dot{H}_f(d))^2\right]^{-1/2} \tag{4.70}$$

On the other hand, from the GRIN lens law we can also obtain the back working distance of the lens for Gaussian illumination, that is, the back focal length measured from the output face of the GRIN lens. The back working distance results when the incident beam has its waist located on the input face of the GRIN lens. At this particular case $d_1 = 0$, $z_R = z_{pR}$, and Eq. (4.68) becomes

$$\begin{aligned} l_g' &= -\frac{H_f(d)\dot{H}_f(d)z_{pR}^2 + H_a(d)\dot{H}_a(d)}{n_0\left[\dot{H}_f^2(d)z_{pR}^2 + \dot{H}_a^2(d)\right]} \\ &= \frac{1}{2n_0 g^2(d)} \frac{d}{dz} \ln\left[\dot{H}_f^2(z)z_{pR}^2 + \dot{H}_a^2(z)\right]\Big|_{z=d} \end{aligned} \tag{4.71}$$

Equation (4.71) reduces to the back working distance for uniform illumination, Eqs. (3.33) or (3.60), when $z_{pR} \to \infty$.

An example of the difference in behavior between Gaussian and uniform beams concerns the image shift [4.17–4.19], i.e., the difference between the position of the image for Gaussian illumination and its position for uniform illumination, as shown in Fig. 4.4. Comparing Eqs. (3.27) and (4.68), we obtain the image shift as

$$\begin{aligned} \Delta d' = d_g' - d_1' &= -\frac{\dot{H}_f(d)z_R^2}{n_0\left[\dot{H}_a(d) + n_0 d_1 \dot{H}_f(d)\right]\left[\dot{H}_f^2(d)z_R^2 + (\dot{H}_a(d) + n_0 d_1 \dot{H}_f(d))^2\right]} \\ &= \frac{1}{2n_0 g^2(d)} \frac{d}{dz}\left\{\ln\left[1 + \frac{\dot{H}_f^2(z)z_R^2}{(\dot{H}_a(z) + n_0 d_1 \dot{H}_f(z))^2}\right]\right\}\Bigg|_{z=d} \end{aligned} \tag{4.72}$$

When the waist of the incident Gaussian beam and the point source for uniform illumination are located at the input face of the GRIN lens, that is, when both sources are located at the input, Eq. (4.72) reduces to

$$\begin{aligned} (\Delta d')_{d_1=0} &= -\frac{\dot{H}_f(d)z_{pR}^2}{n_0 \dot{H}_a(d)\left[\dot{H}_f^2(d)z_{pR}^2 + \dot{H}_a^2(d)\right]} = \frac{z_{pR}^2}{n_0^2 l\left[\dot{H}_f^2(d)z_{pR}^2 + \dot{H}_a^2(d)\right]} \\ &= \frac{1}{2n_0 g^2(d)} \frac{d}{dz}\left\{\ln\left[1 + \frac{\dot{H}_f^2(z)z_{pR}^2}{\dot{H}_a^2(z)}\right]\right\}\Bigg|_{z=d} \end{aligned} \tag{4.73}$$

where l is the front working distance for uniform illumination given by Eq. (3.34).

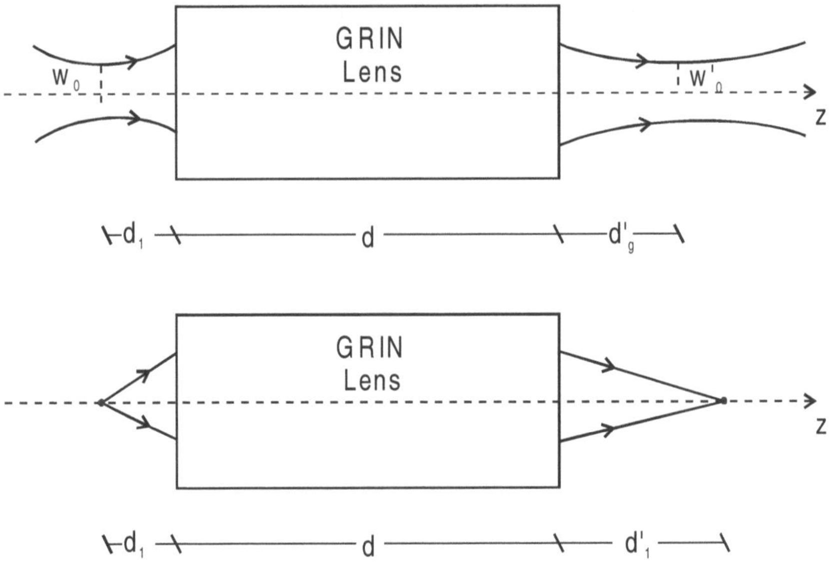

Fig. 4.4. Comparison between Gaussian beam and uniform illumination: image shift.

In the same way, comparing the behavior of a GRIN lens illuminated by a Gaussian plane beam and by a uniform plane wave, the focal shift can be evaluated as shown in Fig. 4.5. From Eqs. (3.10) and (4.71) we can express the focal shift (or the working distance shift) as

$$\begin{aligned}\Delta l' = l'_g - l' &= \frac{\dot{H}_a(d)}{n_0 \dot{H}_f(d)[\dot{H}_f^2(d) z_{pR}^2 + \dot{H}_a^2(d)]} = -\frac{1}{[\dot{H}_f^2(d) z_{pR}^2 + \dot{H}_a^2(d)]} \\ &= \frac{1}{2n_0 g^2(d)} \frac{d}{dz}\left\{ \ln\left[z_{pR}^2 + \frac{\dot{H}_a^2(z)}{\dot{H}_f^2(z)} \right] \right\}\Bigg|_{z=d}\end{aligned} \tag{4.74}$$

By comparing Eqs. (4.73–4.74), we have

$$\frac{(\Delta d')_{d_1=0}}{\Delta l'} = -\left(\frac{z_{pR}}{n_0 l}\right)^2 = -\left(\frac{\pi w_0^2}{l\lambda}\right)^2 \tag{4.75}$$

where Eq. (4.18) has been used.

Thus, the relationship between the image shift at $d_1 = 0$ and the focal shift is proportional to the square of the ratio between the square of the waist of the Gaussian beam and the front working distance of the GRIN lens.

It is essential to know image and focal shifts for optimized performance of devices such as optical GRIN connectors illuminated with Gaussian beams. Any deviation from position of the image of the back focus has the effect of reducing the efficiency of coupling, there by adding insertion losses to the device.

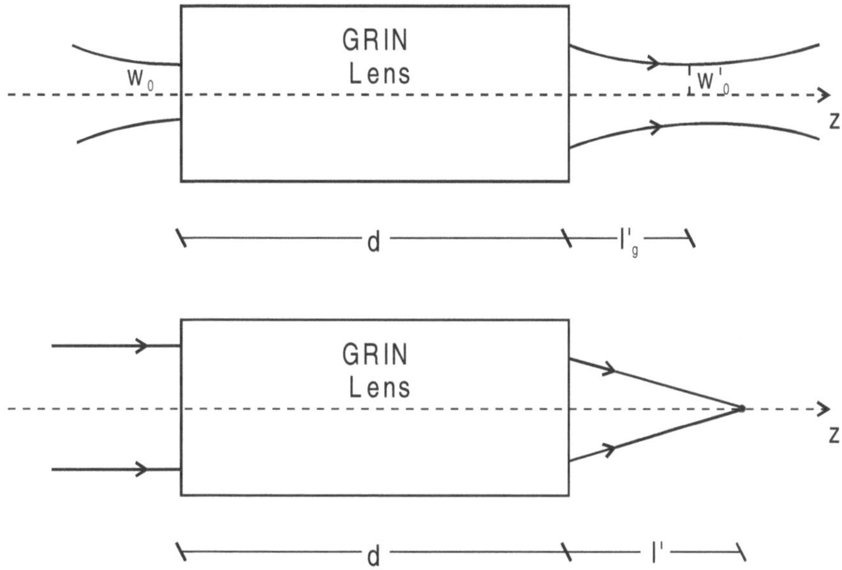

Fig. 4.5. Comparison between Gaussian and uniform plane waves: focal shift.

Another example of the difference in behavior between Gaussian beams and uniform waves occurs when the waist of the incident beam is at the front focal plane of the GRIN lens, in which case the emerging beam has a waist at the back focal plane. The position of the front focal plane measured from the input face of the GRIN lens for uniform illumination is given by Eq. (3.34). When the input waist is located at this plane, $d_1 = l$, the output waist is at the back focal plane, that is

$$d'_g = -\frac{H_f(d)}{n_0 \dot{H}_f(d)} = l' \tag{4.76}$$

where Eqs. (3.33) and (4.68) has been used.

To evaluate image and focal shifts we apply the above results to the selfoc lens, since this kind of lens is used mainly as a GRIN connector in devices for optical communications. In this case, the axial and field rays are given by Eqs. (1.96).

With Eqs. (1.96) inserted into Eqs. (4.72–4.74) and $g(d) = g_0$, image and focal shifts become

$$\Delta d' = \frac{w^4(0)\big/ w_{fm}^4}{n_0 g_0 \sin^2(g_0 d)[\cot an(g_0 d) - n_0 g_0 d_1]\left[\dfrac{w^4(0)}{w_{fm}^4} + (\cot an(g_0 d) - n_0 g_0 d_1)^2\right]} \tag{4.77}$$

$$(\Delta d')_{d_1=0} = \frac{w_0^4 / w_{fm}^4}{n_0 g_0 \mathrm{cotan}(g_0 d)\sin^2(g_0 d)\left[\dfrac{w_0^4}{w_{fm}^4} + \mathrm{cotan}^2(g_0 d)\right]} \quad (4.78)$$

$$\Delta l' = -\frac{1}{n_0 g_0 \tan(g_0 d)\sin^2(g_0 d)\left[\dfrac{w_0^4}{w_{fm}^4} + \mathrm{cotan}^2(g_0 d)\right]} \quad (4.79)$$

where Eqs. (4.14) and (4.18) have been used, and w_{fm} is the waist radius of the fundamental mode in a selfoc lens given by Eq. (4.47).

Note that when $w_{fm} = w(0)$ or $w_{fm} = w_0$, Eqs. (4.77–4.79) reduce to

$$\Delta d' = \frac{1}{n_0 g_0 \sin^2(g_0 d)[\mathrm{cotan}(g_0 d) - n_0 g_0 d_1]\left[1 + (\mathrm{cotan}(g_0 d) - n_0 g_0 d_1)^2\right]} \quad (4.80)$$

$$(\Delta d')_{d_1=0} = \frac{1}{n_0 g_0 \mathrm{cotan}(g_0 d)} = -d'_1|_{d_1=0} \quad (4.81)$$

$$\Delta l' = -\frac{1}{n_0 g_0 \tan(g_0 d)} = -l' \quad (4.82)$$

Equations (4.81–4.82) give the image position and the back working distance for uniform illumination, and they indicate once again the adiffractional behavior of light propagation inside the GRIN lens for $w_{fm} = w_0$.

Figure 4.6a shows the variation of image shift against normalized thickness $g_0 d$ for $w(0) \leq w_{fm}$ and $w(0) \rangle w_{fm}$, and for object distance $d_1 = 100\,\mu m$. Note that when

$$d_1 = \frac{\mathrm{cotan}(g_0 d)}{n_0 g_0} \quad (4.83)$$

that is, when the object distance is equal to the front working distance given by Eq. (3.34) the image shift goes to infinity, since for uniform illumination $d'_1 \to \infty$, and for Gaussian illumination $d'_g = \cos(g_0 d) / n_0 g_0$.

Figure 4.6 depicts focal shift versus $g_0 d$ for $w(0) \leq w_{fm}$ and $w(0) \rangle w_{fm}$. In both figures $\lambda = 1.56\,\mu m$. Selfoc lens data correspond to a w–type selfoc microlens with diameter of 2 mm [4.20].

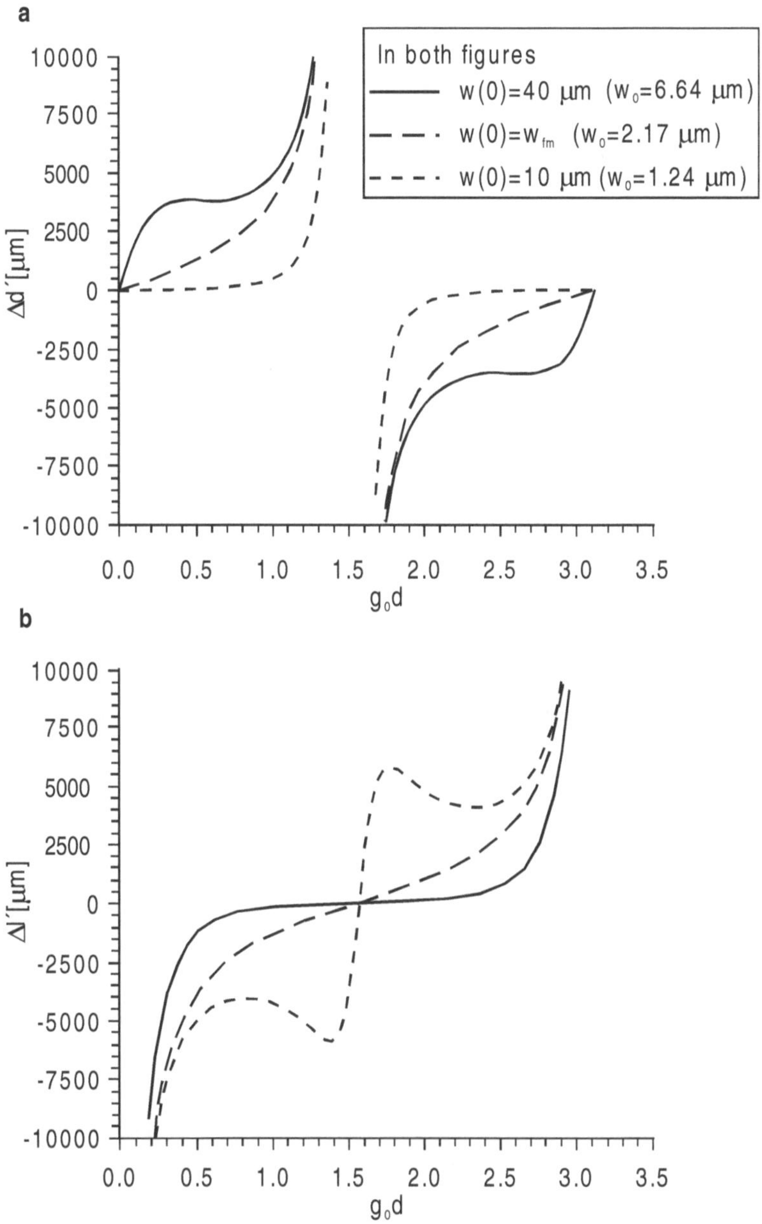

Fig. 4.6. Variation of **a** image and **b** focal shifts with normalized thickness of selfoc lens. Calculations have been made for $d_0 = 100\,\mu m$ (in a), $\lambda = 1.56\,\mu m$, $w_{fm} = 23.04\,\mu m$, $n_0 = 1.59$, and $g_0 = 0.294\,mm^{-1}$.

4.4 Effective Aperture

Until now we have not considered the finite cross–section of the GRIN lens; we shall now study the effect of this limitation on Gaussian beam propagation through a GRIN lens of radius a and thickness d. A spherical Gaussian beam will be confined in a GRIN lens if the following condition is satisfied

$$\frac{w(z)}{|G(z)|_M} \le a \tag{4.84}$$

where $|G(z)|_M$ denotes the maximum value of $|G(z)|$ in the lens.

Equation (4.84) indicates that it will be necessary to define the effective aperture a_e of the input face of the GRIN lens (Fig. 4.7) since not all the Gaussian beam reaching the lens will be confined through it.

From Eq. (4.84) it follows that the effective aperture of the input face of the GRIN lens is given by

$$a_e = \frac{w(0)}{|G(z)|_M} \tag{4.85}$$

where Eq. (4.13) has been used.

The evolution of a_e for ray confinement inside the GRIN lens can be written as

$$a_e(z) = a_e|G(z)| \tag{4.86}$$

verifying

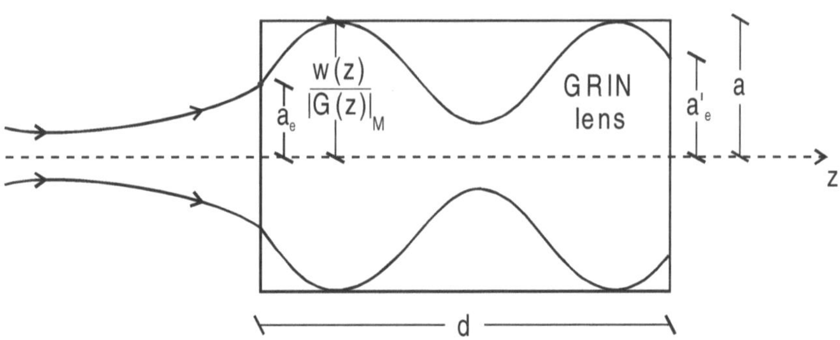

Fig. 4.7. Geometry for evaluating the effective aperture of a GRIN lens illuminated by a Gaussian beam.

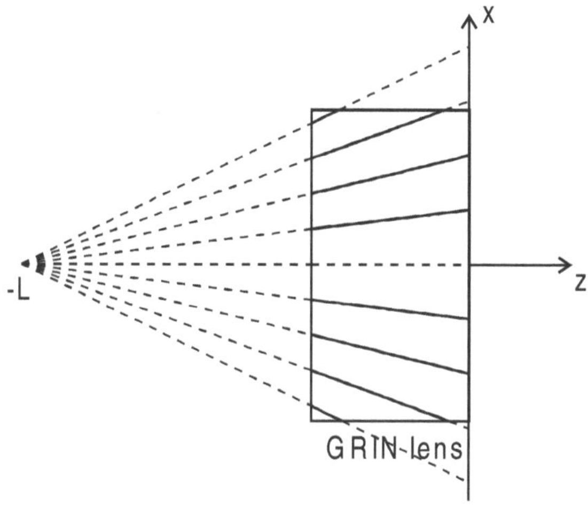

Fig. 4.8. Equi-index cones for a GRIN lens with divergent linear taper function.

$$0 \leq a_e(z) \leq a \tag{4.87}$$

Then the effective aperture for the output face of the GRIN lens will be given by

$$a'_e = a_e|G(d)| \tag{4.88}$$

As mentioned in Sect. 4.2, $|G(z)|_M$ is obtained for $u(z^+)$ as $w_{fm}(0) \geq w(0)$ or $w_{fm}(0) \langle w(0)$ and $R(0) \rangle 0$, and for $u(z^-)$ as $w_{fm}(0) \geq w(0)$ or $w_{fm}(0) \langle w(0)$ and $R(0) \langle 0$.

For the sake of simplicity, we suppose an incident Gaussian beam that has its waist located on the input face of the GRIN lens. In other words, for the lens illuminated by a plane Gaussian beam, $R(0) \to \infty$, $w(0) \to w_0$ and Eqs. (1.92), (4.15), and (4.56–4.60) provide

$$|G_p(z^+)|_M = \frac{g_0 w_{fm}^4(0)}{g(z^+) w_0^4} \quad \text{when} \quad w_{fm}(0) \rangle w_0 \tag{4.89}$$

$$|G_p(z^-)|_M = \frac{g_0}{g(z^-)} \quad \text{when} \quad w_{fm}(0) \langle w_0 \tag{4.90}$$

and

$$|G_p(z)|_M = \frac{g_0}{g(z)} \quad \text{when} \quad w_{fm}(0) = w_0 \tag{4.91}$$

for all values of z inside the GRIN lens.

We can apply these results to a GRIN lens with a divergent linear taper function given by [4.21]

$$g(z) = \frac{g_0}{1 + z/L} \tag{4.92}$$

where g_0 is the value of g(z) at the input face of the GRIN lens, and L is the distance from this face to the common apex of the equi–index cones (Fig. 4.8).

In this kind of GRIN lens, axial and field rays are given by

$$H_a(z) = \frac{1}{g_0}\left(1 + \frac{z}{L}\right)^{1/2} \sin\left[g_0 L \ln\left(1 + \frac{z}{L}\right)\right] \tag{4.93}$$

$$H_f(z) = \left(1 + \frac{z}{L}\right)^{1/2} \cos\left[g_0 L \ln\left(1 + \frac{z}{L}\right)\right] \tag{4.94}$$

For plane Gaussian illumination, the axial positions in which |G| is maximum as $w_{fm}(0) < w_0$ or $w_{fm}(0) > w_0$ can be expressed, respectively, as

$$z_m^+ = L\left[e^{\frac{(2m+1)\pi}{2g_0L}} - 1\right] \tag{4.95}$$

$$z_m^- = L\left[e^{\frac{m\pi}{g_0L}} - 1\right] \tag{4.96}$$

where m is an integer. The first maximum value of G is obtained for m = 0, the second one for m = 1, and so on.

From Eqs. (4.95–4.96) it follows that

$$|G_p(z^+)|_M = \frac{w_{fm}^4(0)}{w_0^4} e^{\frac{(2m+1)\pi}{2g_0L}} \quad \text{when} \quad w_{fm}(0) > w_0 \tag{4.97}$$

$$|G_p(z^-)|_M = e^{\frac{m\pi}{g_0L}} \quad \text{when} \quad w_{fm}(0) < w_0 \tag{4.98}$$

and

$$|G_p(z)|_M = \frac{L + z}{L} \quad \text{when} \quad w_{fm}(0) = w_0 \tag{4.99}$$

for all values of z.

Thus, the evolution of the effective aperture inside the GRIN lens is given by

$$a_e(z) = \frac{w_0^5}{w_{fm}^4(0)} |G(z)| e^{-\frac{(2m+1)\pi}{2g_0L}} \quad \text{when} \quad w_{fm}(0) > w_0 \tag{4.100}$$

$$a_e(z) = w_0 |G(z)| e^{-\frac{m\pi}{g_0L}} \quad \text{when} \quad w_{fm}(0) < w_0 \tag{4.101}$$

and

$$a_e(z) = \frac{L}{L+z} w_{fm}(z) \quad \text{when} \quad w_{fm}(0) = w_0 \tag{4.102}$$

Finally, the results obtained in this chapter can be extended to GRIN lenses illuminated by Gaussian beams with two complex curvatures (elliptical Gaussian beams). Likewise, the diffractive effect due to finite aperture (diffraction–limited propagation of light) can also be evaluated, in the same way as in Chap. 3, by Lommel functions with complex arguments [4.22].

5 GRIN Media with Loss or Gain

5.1 Introduction

In the previous chapters, we discussed light propagation through a passive GRIN medium that exhibited neither loss or gain. This is an idealization, since most dielectric media have losses in the optical domain spectrum. Likewise, the active materials in lasers or waveguides with high amplification can be considered dielectric media exhibiting gain rather than loss [5.1–5.10].

Effects of gain or loss in GRIN materials can be phenomenologically taken into account by using a complex refractive index in the Helmholtz or parabolic wave equations (1.25) and (1.29). In other words, propagation of light beams in active GRIN media can be studied from expressions for passive GRIN media, if we employ formally a complex refractive index instead of the real one.

This chapter examines the implications of using a complex index on light propagation in GRIN materials, in order to study how Gaussian beams may be transformed into uniform beams and vice versa by active GRIN materials [5.11–5.16].

5.2 Active GRIN Materials: Complex Refractive Index

We consider an active GRIN medium with rotational symmetry around the z–axis, limited by plane parallel faces of thickness d perpendicular to the axis surrounded by a vacuum, and whose refractive index can be expressed as in Eq. (1.69b) by

$$n(x,y,z) = n_0\left[1 - \frac{g^2(z)}{2}(x^2 + y^2)\right] \tag{5.1}$$

where n_0 and $g(z)$ are now the complex refractive index along the z–axis and the complex gradient parameter, respectively.

We assume that n_0 is given by

$$n_0 = n_{0r} + i n_{0i} \tag{5.2}$$

where n_{0r} and n_{0i} are real constants.

Similarly, we introduce the complex function

$$g(z) = g_r(z) + i g_i(z) \tag{5.3}$$

where g_r and g_i are real functions of z, whose signs are preserved along the axis.

From Eqs. (5.2–5.3), the real and imaginary parts of n can be written as

$$\mathrm{Re}[n(x,y,z)] = n_{0r} - \left\{\frac{n_{0r}}{2}[g_r^2(z) - g_i^2(z)] - n_{0i} g_r(z) g_i(z)\right\}(x^2+y^2) \tag{5.4a}$$

$$\mathrm{Im}[n(x,y,z)] = n_{0i} - \left\{\frac{n_{0i}}{2}[g_r^2(z) - g_i^2(z)] + n_{0r} g_r(z) g_i(z)\right\}(x^2+y^2) \tag{5.4b}$$

Using these equations, we can classify active media according to the real and imaginary parts of their complex refractive indices.

For the real part of the refractive index, if

$$\frac{g_r^2(z) - g_i^2(z)}{2 g_r(z) g_i(z)} \rangle \frac{n_{0i}}{n_{0r}} \tag{5.5}$$

Re[n] decreases along transverse direction, and the medium has a normal guidance behavior.

An anomalous guidance behavior occurs if

$$\frac{g_r^2(z) - g_i^2(z)}{2 g_r(z) g_i(z)} \langle \frac{n_{0i}}{n_{0r}} \tag{5.6}$$

since Re[n] increases along the transverse direction.

On the other hand, gain or loss is determined by the sign of the imaginary part of n. The medium has loss if

$$\mathrm{Im}[n] \rangle 0 \tag{5.7}$$

and it experiences gain if

$$\mathrm{Im}[n] \langle 0 \tag{5.8}$$

From Eq. (5.7) it follows that the medium suffers on–axis loss ($n_{0i} \rangle 0$), and if

$$\frac{g_r^2(z) - g_i^2(z)}{2 g_r(z) g_i(z)} \rangle -\frac{n_{0r}}{n_{0i}} \tag{5.9}$$

loss decreases with increasing values along the radial direction, turning to gain for transverse distance sufficiently large, that is

$$x^2 + y^2 \rangle \frac{2 n_{0i}}{n_{0i}[g_r^2(z) - g_i^2(z)] + 2 n_{0r} g_r(z) g_i(z)} \tag{5.10}$$

This anomalous behavior explains why light propagation in this type of active medium is unstable.

In the same way, if

$$\frac{g_r^2(z)-g_i^2(z)}{2g_r(z)g_i(z)}\langle -\frac{n_{0r}}{n_{0i}} \tag{5.11}$$

loss increases with distance from the axis. This means that stable propagation of light occurs.

From Eq. (5.8) it follows that the medium has on–axis gain ($n_{0i}\langle 0$), and if Eq. (5.9) is fulfilled, gain increases along transverse direction (the unstable case). On the contrary, if Eq. (5.11) is verified, gain decreases along the transverse direction and turns to loss for distance given by Eq. (5.10). This is the stable case [5.3].

5.3 The Kernel Function

We consider the propagation of a monochromatic field ψ of wavelength λ, in an active GRIN medium satisfying the parabolic wave equation (2.58) for complex refractive index

$$2ik(n_{0r}+in_{0i})\frac{\partial\psi}{\partial z}=H\psi \tag{5.12}$$

where H is the Hamiltonian operator given by

$$H=k^2(x^2+y^2)(n_{0r}+in_{0i})^2[g_r(z)+ig_i(z)]^2-\nabla_\perp^2 \tag{5.13}$$

In this case, the Hamiltonian becomes a non–Hermitian operator.

Equation (5.12) describes the evolution of the field ψ as a function of z through an active GRIN medium. The kernel of Eq. (5.12) can be derived from Eq. (2.21) by using a complex refractive index and the complex axial and field ray solutions of the paraxial complex ray equation [5.11]

$$\ddot{H}_{\binom{a}{f}}(z)+g^2(z)H_{\binom{a}{f}}=0 \tag{5.14}$$

resulting from the paraxial ray equation (1.70b) by means of analytic continuation.

Therefore, the two independent solutions of Eq. (5.14) can be obtained from axial and field rays for a passive GRIN medium if we use a complex gradient parameter. With the aid of these two complex rays we can express any other paraxial complex ray as a linear combination of both rays.

In this way, the position and slope of the axial and field rays are given by

$$\overset{(\cdot)}{H}_a(z)=\overset{(\cdot)}{H}_{ar}(z)+i\overset{(\cdot)}{H}_{ai}(z) \tag{5.15a}$$

$$\overset{(\cdot)}{H}_{f}(z)=\overset{(\cdot)}{H}_{fr}(z)+i\overset{(\cdot)}{H}_{fi}(z) \tag{5.15b}$$

where $\overset{(\cdot)}{H}_{ar}$, $\overset{(\cdot)}{H}_{fr}$ and $\overset{(\cdot)}{H}_{ai}$, $\overset{(\cdot)}{H}_{fi}$ are the real and imaginary parts of $\overset{(\cdot)}{H}_{a}$ and $\overset{(\cdot)}{H}_{f}$, respectively.

Both rays are defined with initial conditions (1.72) and (1.74) and satisfy the Lagrange's invariant

$$\dot{H}_a(z)H_f(z)-H_a(z)\dot{H}_f(z)=1 \tag{5.16}$$

that is

$$\dot{H}_{ar}(z)H_{fr}(z)+H_{ai}(z)\dot{H}_{fi}(z)-\left[\dot{H}_{ai}(z)H_{fi}(z)+H_{ar}(z)\dot{H}_{fr}(z)\right]=1 \tag{5.17a}$$

$$\dot{H}_{ar}(z)H_{fi}(z)+\dot{H}_{ai}(z)H_{fr}(z)=H_{ar}(z)\dot{H}_{fi}(z)+H_{ai}(z)\dot{H}_{fr}(z) \tag{5.17b}$$

On the other hand, the kernel function at any plane inside the active GRIN medium can be expressed from Eq. (2.21) by the real and imaginary parts of the complex refractive index and axial and field rays as

$$\begin{aligned}K(x,y,x_0,y_0;z)=&\frac{k\sqrt{\left[n_{0r}H_{ar}(z)+n_{0i}H_{ai}(z)\right]^2+\left[n_{0i}H_{ar}(z)-n_{0r}H_{ai}(z)\right]^2}}{i2\pi\left|H_a(z)\right|^2}\\&\exp\{ikn_{0r}z\}\exp\{-kn_{0i}z\}\exp\left\{i\tan^{-1}\left[\frac{n_{0i}H_{ar}(z)-n_{0r}H_{ai}(z)}{n_{0r}H_{ar}(z)+n_{0i}H_{ai}(z)}\right]\right\}\\&\exp\left\{-\frac{k}{2}\theta(x,y,x_0,y_0;z)\right\}\exp\left\{i\frac{k}{2}\Omega(x,y,x_0,y_0;z)\right\}\end{aligned} \tag{5.18}$$

where

$$\begin{aligned}\theta(x,y,x_0,y_0;z)=&\frac{1}{\left|H_a(z)\right|^2}\Big\{\Big[n_{0r}\left(H_{ar}(z)\dot{H}_{ai}(z)-H_{ai}(z)\dot{H}_{ar}(z)\right)\\&+n_{0i}\left(H_{ai}(z)\dot{H}_{ai}(z)+H_{ar}(z)\dot{H}_{ar}(z)\right)\Big]\left(x^2+y^2\right)\\&+\Big[n_{0r}\left(H_{ar}(z)H_{fi}(z)-H_{ai}(z)H_{fr}(z)\right)\\&+n_{0i}\left(H_{ai}(z)H_{fi}(z)+H_{ar}(z)H_{fr}(z)\right)\Big]\left(x_0^2+y_0^2\right)\\&-2\left[n_{0i}H_{ar}(z)-n_{0r}H_{ai}(z)\right]\left(xx_0+yy_0\right)\Big\}\end{aligned} \tag{5.19}$$

$$\begin{aligned}\Omega(x,y,x_0,y_0;z)=&\frac{1}{\left|H_a(z)\right|^2}\Big\{\Big[n_{0r}\left(H_{ar}(z)\dot{H}_{ar}(z)+H_{ai}(z)\dot{H}_{ai}(z)\right)\\&+n_{0i}\left(H_{ai}(z)\dot{H}_{ar}(z)-H_{ar}(z)\dot{H}_{ai}(z)\right)\Big]\left(x^2+y^2\right)\\&+\Big[n_{0r}\left(H_{ar}(z)H_{fr}(z)+H_{ai}(z)H_{fi}(z)\right)\\&+n_{0i}\left(H_{ai}(z)H_{fr}(z)-H_{ar}(z)H_{fi}(z)\right)\Big]\left(x_0^2+y_0^2\right)\\&-2\left[n_{0r}H_{ar}(z)+n_{0i}H_{ai}(z)\right]\left(xx_0+yy_0\right)\Big\}\end{aligned} \tag{5.20}$$

and

$$|H_a(z)| = \sqrt{H_{ar}^2(z) + H_{ai}^2(z)} \tag{5.21}$$

Equation (5.18) represents the paraxial approximation to the spread function of an active GRIN medium. For this case, the medium has on–axis gain or loss given by $\exp\{-kn_{0i}z\}$, and the amplitude of the kernel is modulated by a Gaussian function. The analysis becomes simpler if we note that the amplitude factor $\exp\left\{-\frac{k}{2}\theta\right\}$ of Eq. (5.18), for an on–axis point source located at the input plane $z=0$, reduces to

$$\exp\left\{-\frac{x^2+y^2}{w^2(z)}\right\} \tag{5.22}$$

where the spot size of the Gaussian function is given by

$$w^2(z) = \frac{\lambda H_a^2(z)}{\theta|_{x_0=y_0=0}} \tag{5.23}$$

Then, an active GRIN medium works like a quadratic phase transformer with a Gaussian mask induced by the complex refractive index.

Finally, in the lossless case, the imaginary parts of $\overset{(\cdot)}{H}_a$, and $\overset{(\cdot)}{H}_f$, n_0, and g vanish; θ tends to zero; and Eq. (5.18) becomes Eq. (2.21), as expected.

5.4 Focal Distance and Focal Shift for Uniform Illumination

We will study light propagation in an active GRIN medium by the transmittance function for uniform illumination in order to evaluate the focal distance and the focal shift. We now suppose an active GRIN medium of thickness d illuminated by a monochromatic uniform plane wave of unit amplitude and wavelength λ. The transmittance function given by Eq. (3.8) can be written by means of analytic continuation for an active medium as

$$t(x,y;d) = \frac{\exp\{i\varphi(d)\}\exp\{-kn_{0i}d\}}{|H_f(d)|}\exp\left\{i\frac{kR'(d)}{2}\left(x^2+y^2\right)\right\}\exp\left\{-\frac{kI'(d)}{2}\left(x^2+y^2\right)\right\} \tag{5.24}$$

where φ is the on–axis phase

$$\varphi(d) = kn_{0r}d - \tan^{-1}\left[\frac{H_{fi}(d)}{H_{fr}(d)}\right] \tag{5.25}$$

R' and I' are the real and imaginary parts of $n_0\dot{H}_f(d)/H_f(d)$, that is

$$R'(d) = \mathrm{Re}\left[\frac{n_0\dot{H}_f(d)}{H_f(d)}\right] = \frac{n_{0r}\left[\dot{H}_{fr}(d)H_{fr}(d) + \dot{H}_{fi}(d)H_{fi}(d)\right] + n_{0i}\left[\dot{H}_{fr}(d)H_{fi}(d) - \dot{H}_{fi}(d)H_{fr}(d)\right]}{|H_f(d)|^2} \tag{5.26a}$$

$$I'(d) = \mathrm{Im}\left[\frac{n_0\dot{H}_f(d)}{H_f(d)}\right] = \frac{n_{0r}\left[\dot{H}_{fi}(d)H_{fr}(d) - \dot{H}_{fr}(d)H_{fi}(d)\right] + n_{0i}\left[\dot{H}_{fr}(d)H_{fr}(d) + \dot{H}_{fi}(d)H_{fi}(d)\right]}{|H_f(d)|^2} \tag{5.26b}$$

and

$$|H_f(d)| = \sqrt{H_{fr}^2(d) + H_{fi}^2(d)} \tag{5.26c}$$

Equation (5.24) represents the complex amplitude distribution at the output plane of the active GRIN medium when it is illuminated by a uniform plane wave. Equation (5.24) may be regarded as the complex amplitude distribution at the output face of a passive GRIN medium with a Gaussian mask located at this plane, whose on–axis value is given by $\exp\{-kn_{0i}d\}$ and of spot size [5.17–5.18]

$$w'(d) = \sqrt{\frac{\lambda}{\pi I'(d)}} \tag{5.27}$$

Next, we evaluate the focal distance of the active GRIN medium. The problem reduces to the calculation of the complex amplitude distribution on a plane located at distance d' from the output face of the active medium in order to find the axial distance for which the irradiance achieves an extrem value. Neglecting the finite dimension of the active medium in a first approach, the complex amplitude distribution at this plane is expressed as

$$\psi(x_1', y_1'; d') = \int_{\Re^2} K^{FP}(x, y, x_1', y_1'; d')t(x, y; d)dxdy \tag{5.28}$$

where

$$K^{FP}(x, y, x_1', y_1'; d') = \frac{1}{id'\lambda}\exp\left\{i\frac{k}{2d'}\left[(x_1' - x)^2 + (y_1' - y)^2\right]\right\} \tag{5.29}$$

is the impulse response over the distance d' in free space.

Substituting Eqs. (5.24) and (5.29) into Eq. (5.28) and integrating, we obtain

$$\psi(x_1', y_1'; d_1') = \frac{\exp\{i\varphi'(d;d')\}\exp\{-kn_{0i}d\}}{d'|H_f(d)||J'(d,d')|}\exp\left\{i\frac{k}{2d'}\left[1-\frac{Q'(d,d')}{d'|J'(d,d')|^2}\right](x_1'^2+y_1'^2)\right\}$$

$$\exp\left\{-\frac{kI'(d)}{2d'^2|J'(d,d')|^2}(x_1'^2+y_1'^2)\right\} \tag{5.30}$$

where

$$Q'(d,d') = R'(d) + \frac{1}{d'} \tag{5.31a}$$

$$|J'(d,d')| = \sqrt{Q'^2(d,d') + I'^2(d)} \tag{5.31b}$$

and

$$\varphi'(d,d') = \varphi(d) - \tan^{-1}\left[\frac{I'(d)}{Q'(d,d')}\right] \tag{5.31c}$$

From Eq. (5.30) it follows that at the observation plane, we have a complex amplitude distribution corresponding to a Gaussian beam whose curvature and beam half–width are given by

$$C(d') = \frac{1}{d'}\left[1 - \frac{Q'(d,d')}{d'|J'(d,d')|^2}\right] \tag{5.32}$$

and

$$w(d') = d'|J'(d,d')|\sqrt{\frac{\lambda}{\pi I'(d)}} = d|J'(d,d')|w'(d) \tag{5.33}$$

We are now interested in that axial length d_a' where the beam half–width is an extremum (Fig. 5.1). The length d_a' can be obtained by the vanishing of the beam curvature. From Eq. (5.32), this condition provides

$$d_a' = \frac{Q'(d,d_a')}{|J'(d,d_a')|^2} \tag{5.34}$$

Equation (5.34) can be written as

$$d_a' = -\frac{R'(d)}{R'^2(d) + I'^2(d)} \tag{5.35}$$

where Eqs. (5.31a–b) have been used.

The length d_a' is known as the back focal distance of the active GRIN medium measured from the output face. At this length, the beam half–width achieves an extremum value and becomes the beam waist. Thus, inserting Eq. (5.35) into Eq. (5.33) and taking into account Eq. (5.31b), the beam waist is given by [5.19]

$$w(d'_a) = \sqrt{\frac{\lambda I'(d)}{\pi[R'^2(d) + I'^2(d)]}} \tag{5.36}$$

that can be expressed in terms of the spot size of the Gaussian mask as

$$w(d'_a) = \frac{w'(d)}{\sqrt{1 + \frac{\pi^2 R'^2(d) w'^4(d)}{\lambda^2}}} \tag{5.37}$$

Likewise, substituting Eq. (5.35) into Eq. (5.30), the irradiance at the back focus is written as

$$|\psi(x'_1, y'_1; d'_a)|^2 = \frac{w'^2(d)}{|H_f(d)|^2 w^2(d'_a)} \exp\{-2kn_{0i}d\} \exp\left\{-\frac{2}{w^2(d'_a)}(x'^2_1 + y'^2_1)\right\} \tag{5.38}$$

Note that for a passive GRIN medium, $I'(d)$ vanishes, then Eq. (5.35) becomes Eq. (3.14), and Eq. (5.38) tends to a two–dimensional delta function at the back focus.

On the other hand, from Eqs. (5.24) and (3.9) it follows that at a distance d'_{ae} from the output face of the active medium such that

$$d'_{ae} = -\frac{1}{R'(d)} \tag{5.39}$$

the complex amplitude distribution is a two–dimensional delta function when a Gaussian mask of spot size

$$w(d) = iw'(d) \tag{5.40}$$

is located at the output face, when the medium is illuminated from the left by a uniform plane wave (Fig. 5.1). In this case, a uniform spherical wave at the output face is obtained.

The difference between d'_a and d'_{ae} defines the focal shift due to gain or loss in the active GRIN medium [5.20–5.21] and is given by

$$\Delta d'_a = d'_a - d'_{ae} = -\frac{d'_{ae}}{1 + \frac{\pi^2 w'^4(d)}{d'^2_{ae}\lambda^2}} \tag{5.41}$$

where Eqs. (5.27), (5.35), and (5.39) have been used.

When Eq. (5.41) is taken into account, Eq. (5.37) can be expressed as

$$\frac{w^2(d'_a)}{w'^2(d)} = -\frac{\Delta d'_a}{d'_{ae}} \tag{5.42}$$

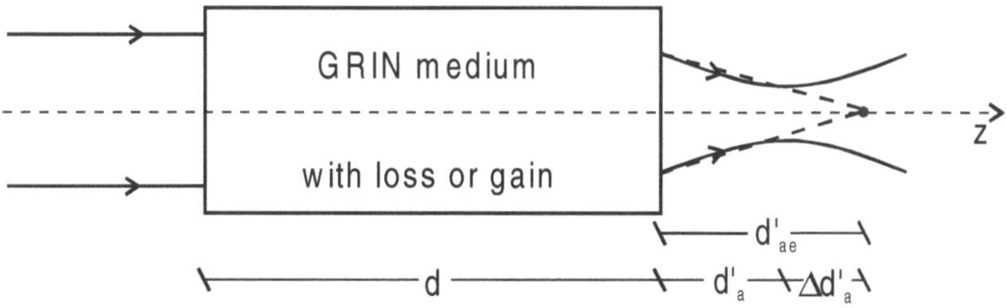

Fig. 5.1. Active GRIN medium illuminated by uniform plane wave; d'_a is the back focal distance for the output Gaussian beam and d'_{ae} is the back focal distance for the output uniform spherical wave when a Gaussian aperture is located at the output face of the active GRIN medium.

To evaluate focal distance, focal shift, beam waist, and spot size of the Gaussian mask, we apply these results to an active selfoc medium, since this kind of medium can be used to optimizes performance of active devices. For our particular case, the complex gradient parameter is given by

$$g(z) = g_0 = g_{0r} + ig_{0i} \tag{5.43}$$

where g_{0r} and g_{0i} are real constants.

The real and imaginary parts of the position and slope of the complex axial and field rays can be obtained from Eqs. (1.96) by means of analytic continuation and are written as

$$H_{ar}(z) = \frac{1}{g_{0r}^2 + g_{0i}^2}\{g_{0r}\sin(g_{0r}z)\cosh(g_{0i}z) + g_{0i}\cos(g_{0r}z)\sinh(g_{0i}z)\} \tag{5.44a}$$

$$H_{ai}(z) = \frac{1}{g_{0r}^2 + g_{0i}^2}\{g_{0r}\cos(g_{0r}z)\sinh(g_{0i}z) - g_{0i}\sin(g_{0r}z)\cosh(g_{0i}z)\} \tag{5.44b}$$

$$H_{fr}(z) = \dot{H}_{ar}(z) = \cos(g_{0r}z)\cosh(g_{0i}z) \tag{5.44c}$$

$$H_{fi}(z) = \dot{H}_{ai}(z) = -\sin(g_{0r}z)\sinh(g_{0i}z) \tag{5.44d}$$

and

$$\dot{H}_{fr}(z) = -g_{0r}\sin(g_{0r}z)\cosh(g_{0i}z) + g_{0i}\cos(g_{0r}z)\sinh(g_{0i}z) \tag{5.44e}$$

$$\dot{H}_{fi}(z) = -[g_{0i}\sin(g_{0r}z)\cosh(g_{0i}z) + g_{0r}\cos(g_{0r}z)\sinh(g_{0i}z)] \tag{5.44f}$$

Substituting Eqs. (5.44) into Eqs. (5.26a–b), we get

$$R'(d) = n_{0r}\frac{\left[-g_{0r}\sin(g_{0r}d)\cos(g_{0r}d) + g_{0i}\sinh(g_{0i}d)\cosh(g_{0i}d)\right]}{\cos^2(g_{0r}d)\cosh^2(g_{0i}d) + \sin^2(g_{0r}d)\sinh^2(g_{0i}d)} + n_{0i}\frac{\left[g_{0r}\sinh(g_{0i}d)\cosh(g_{0i}d) + g_{0i}\sin(g_{0r}d)\cos(g_{0r}d)\right]}{\cos^2(g_{0r}d)\cosh^2(g_{0i}d) + \sin^2(g_{0r}d)\sinh^2(g_{0i}d)} \tag{5.45a}$$

$$I'(d) = -n_{0r}\frac{\left[g_{0r}\sinh(g_{0i}d)\cosh(g_{0i}d) + g_{0i}\sin(g_{0r}d)\cos(g_{0r}d)\right]}{\cos^2(g_{0r}d)\cosh^2(g_{0i}d) + \sin^2(g_{0r}d)\sinh^2(g_{0i}d)} + n_{0i}\frac{\left[-g_{0r}\sin(g_{0r}d)\cos(g_{0r}d) + g_{0i}\sinh(g_{0i}d)\cosh(g_{0i}d)\right]}{\cos^2(g_{0r}d)\cosh^2(g_{0i}d) + \sin^2(g_{0r}d)\sinh^2(g_{0i}d)} \tag{5.45b}$$

By using Eqs. (5.45) focal distance, focal shift, beam waist, and spot size of the equivalent Gaussian mask can be evaluated. Figures 5.2 and 5.3 depict $w^2(d'_a)$ and $w'^2(d)$ versus thickness of the active selfoc medium. Unstable (Fig. 5.2) and stable (Fig. 5.3) cases are considered. Calculations have been made for $n_{0r} = 1.6$, $n_{0i} = 0.1$, $\lambda = 1.3\,\mu m$, $g_{0r} = 0.2\,mm^{-1}$, $g_{0i} = 0.003\,mm^{-1}$ (unstable case) and $g_{0i} = -0.003\,mm^{-1}$ (stable case). The medium suffers on–axis loss. The stable case involves power concentration, and in the unstable case, power spread occurs. In the latter case, there is a small region in which power is concentrated. Note that when $w'(d) \to \infty$, $w(d'_a) \to 0$ and the irradiance at the back focus tends to a two–dimensional delta function.

Finally, the front focal distance measured from the input face of the active GRIN medium can be calculated when the medium is illuminated from the right by a uniform plane wave.

The front focal distance is given by

$$d_a = \frac{Re(d)}{Re^2(d) + I^2(d)} \tag{5.46}$$

where Re and I are the real and imaginary parts of $-n_0 \dot{H}_f / \dot{H}_a$, that is

$$Re(d) = Re\left[-\frac{n_0\dot{H}_f(d)}{\dot{H}_a(d)}\right] \tag{5.47a}$$

$$= -\frac{n_{0r}\left[\dot{H}_{fr}(d)\dot{H}_{ar}(d) + \dot{H}_{fi}(d)\dot{H}_{ai}(d)\right] + n_{0i}\left[\dot{H}_{fr}(d)\dot{H}_{ai}(d) - \dot{H}_{fi}(d)\dot{H}_{ar}(d)\right]}{\dot{H}_{ar}^2(d) + \dot{H}_{ai}^2(d)}$$

$$I(d) = Im\left[-\frac{n_0\dot{H}_f(d)}{\dot{H}_a(d)}\right] \tag{5.47b}$$

$$= -\frac{n_{0r}\left[\dot{H}_{fi}(d)\dot{H}_{ar}(d) - \dot{H}_{fr}(d)\dot{H}_{ai}(d)\right] + n_{0i}\left[\dot{H}_{fr}(d)\dot{H}_{ar}(d) + \dot{H}_{fi}(d)\dot{H}_{ai}(d)\right]}{\dot{H}_{ar}^2(d) + \dot{H}_{ai}^2(d)}$$

and the waist of the emerging Gaussian beam can be expressed as

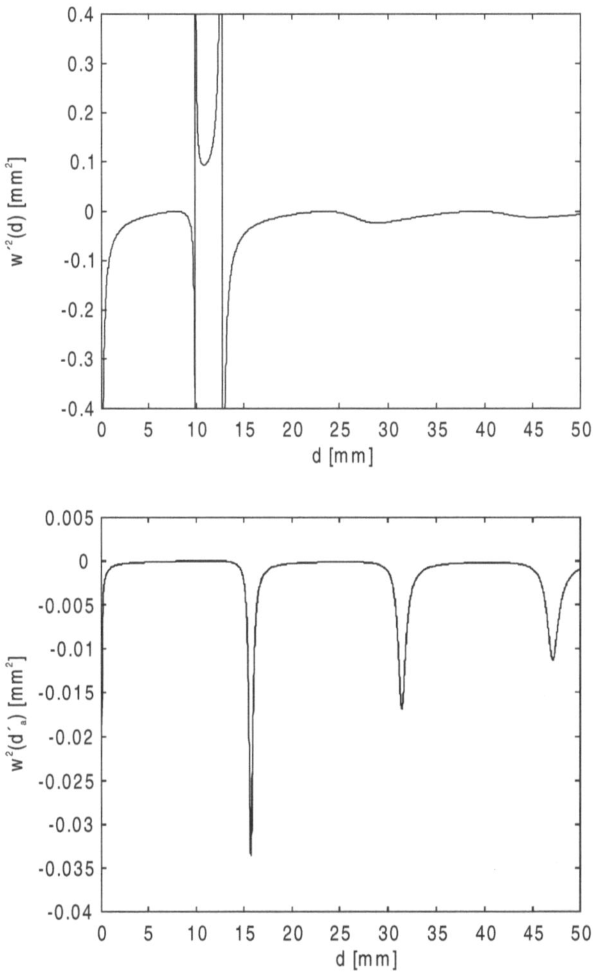

Fig. 5.2. The beam waist $w^2(d'_a)$ and the spot size of the Gaussian mask $w'^2(d)$ versus thickness of active GRIN medium. Calculations have been made for $n_{0r} = 1.6$, $n_{0i} = 0.1$, $\lambda = 1.3\,\mu m$, $g_{0r} = 0.2\,mm^{-1}$ and $g_{0i} = 0.003\,mm^{-1}$ (unstable case).

$$w(d_a) = \sqrt{-\frac{\lambda I(d)}{\pi[Re^2(d) + I^2(d)]}} \tag{5.48}$$

Thus, if we illuminate the active GRIN medium with a Gaussian beam from a source with a waist $w(d_a)$ located at d_a, we obtain uniform plane illumination at the output face of the medium. This property indicates that the active GRIN medium works like a passive GRIN medium with a Gaussian mask of spot size

$$w'_{up}(d) = \sqrt{-\frac{\lambda}{\pi I(d)}} \tag{5.49}$$

to produce uniform illumination, as is discussed in Sect. 5.6.

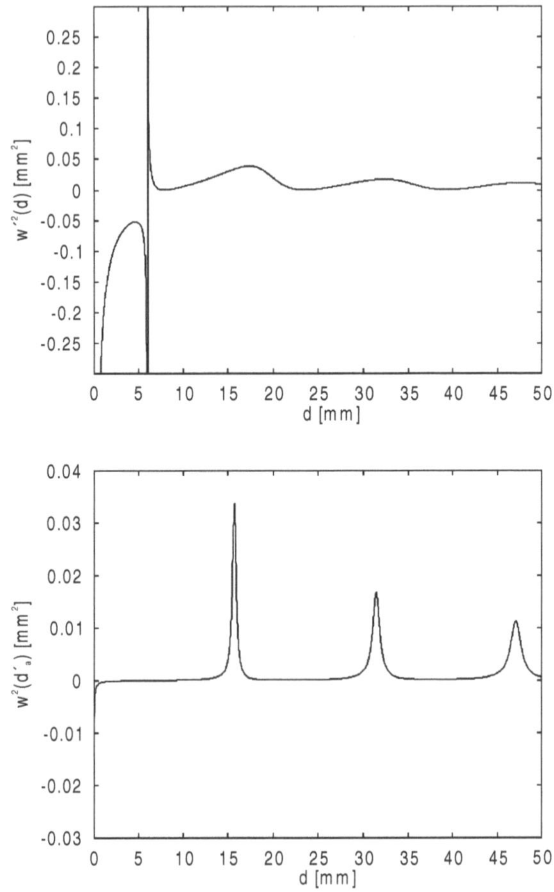

Fig. 5.3. $w^2(d'_a)$ and $w'^2(d)$ versus thickness of active GRIN medium. Calculations have been made for $n_{0r} = 1.6$, $n_{0i} = 0.1$, $\lambda = 1.3\,\mu m$, $g_{0r} = 0.2\,mm^{-1}$ and $g_{0i} = -0.003\,mm^{-1}$ (stable case).

5.5 Gaussian Illumination in an Active GRIN Medium: Beam Parameters

When an active GRIN medium is illuminated, for instance, by a spherical Gaussian beam, the complex amplitude distribution at the input face is given by Eq. (4.1). Then, the field at the output face of an active medium can be written from Eq. (4.7) by analytic continuation as

$$\psi(x,y;d)=\frac{w_0}{w(0)G(d)}\exp\{i\varphi(d)\}\exp\left\{i\frac{\pi U(d)}{\lambda}\left(x^2+y^2\right)\right\} \tag{5.50}$$

where U(d) is the complex curvature of the output beam

$$U(d)=n_0\frac{d}{dz}\ln G(z)|_{z=d}=n_0\dot{G}(d)G^{-1}(d) \tag{5.51}$$

$\varphi(d)$ is the on–axis phase

$$\varphi(d)=\varphi(0)+kn_0 d \tag{5.52}$$

$\varphi(0)$ is given by Eq. (4.3), where n_0 is the complex refractive index along the z–axis.

Likewise, real and imaginary parts of G and $\dot{G}(d)$ can now be expressed as

$$\overset{(\cdot)}{G}_r(d)=\dot{H}_{fr}(d)+a\overset{(\cdot)}{H}_{ar}(d)+b\overset{(\cdot)}{H}_{ai}(d) \tag{5.53a}$$

$$\overset{(\cdot)}{G}_i(d)=\dot{H}_{fi}(d)-b\overset{(\cdot)}{H}_{ar}(d)+a\overset{(\cdot)}{H}_{ai}(d) \tag{5.53b}$$

where

$$a=\frac{1}{|n_0|^2}\left[\frac{n_{0r}}{R(0)}+\frac{n_{0i}\lambda}{\pi w^2(0)}\right] \tag{5.54a}$$

$$b=\frac{1}{|n_0|^2}\left[\frac{n_{0i}}{R(0)}-\frac{n_{0r}\lambda}{\pi w^2(0)}\right] \tag{5.54b}$$

and Eqs. (4.2), (4.10), and (5.15) have been used. R(0) and w(0) are the radius of curvature and the beam half–width at the input face of the active medium, respectively.

On the other hand, the complex curvature defining the field is related to the complex ray q by Eq. (4.30), and can be rewritten for an active medium as [5.22–5.25]

$$U(d) = n_0\dot{q}(d)q^{-1}(d) = \frac{1}{|q(d)|^2}\{n_{0r}[q_r(d)\dot{q}_r(d) + q_i(d)\dot{q}_i(d)]$$
$$+ n_{0i}[\dot{q}_r(d)q_i(d) - q_r(d)\dot{q}_i(d)] +$$
$$+ i[n_{0r}[q_r(d)\dot{q}_i(d) - \dot{q}_r(d)q_i(d)] + n_{0i}[q_r(d)\dot{q}_r(d) + q_i(d)\dot{q}_i(d)]]\} \tag{5.55}$$

where Eqs. (4.29) and (4.26a) have been used, that is

$$\overset{(\cdot)}{q}_r(d) = w_0\left[\overset{(\cdot)}{G}_r(d) - c\overset{(\cdot)}{G}_i(d)\right] \tag{5.56a}$$

$$\overset{(\cdot)}{q}_i(d) = w_0\left[\overset{(\cdot)}{G}_i(d) + c\overset{(\cdot)}{G}_r(d)\right] \tag{5.56b}$$

and

$$c = \frac{\lambda d_1}{\pi w_0^2} \tag{5.56c}$$

where w_0 is the waist of the incident Gaussian beam located at distance d_1 from the input face of the active GRIN medium.

From Eqs. (5.56) it follows that

$$|q(d)|^2 = w^2(0)|G(d)|^2; \quad |\dot{q}(d)|^2 = w^2(0)|\dot{G}(d)|^2 \tag{5.57}$$

where Eqs. (4.4–4.5) have been used.

Equation (5.57) indicates the evolution of the modulus of the position and slope of the complex ray in terms of G and $\dot{G}$ functions, respectively. When Eqs. (5.55–5.57) are taken into account, Eq. (5.50) becomes

$$\psi(x, y; d) = \frac{w_0}{q(d)}\exp\{ikn_0 d\}\exp\left\{i\frac{\pi}{\lambda}U(d)[x^2 + y^2]\right\} \tag{5.58}$$

Comparing Eq. (4.11) with Eq. (5.55), the half–width and the radius of curvature of the output Gaussian beam in terms of q and $\dot{q}$ are given by

$$w^2(d) = \frac{\lambda|q(d)|^2}{\pi\{n_{0r}[q_r(d)\dot{q}_i(d) - \dot{q}_r(d)q_i(d)] + n_{0i}[q_r(d)\dot{q}_r(d) + q_i(d)\dot{q}_i(d)]\}} \tag{5.59a}$$

$$R(d) = \frac{|q(d)|^2}{n_{0r}[q_r(d)\dot{q}_r(d) + q_i(d)\dot{q}_i(d)] + n_{0i}[\dot{q}_r(d)q_i(d) - q_r(d)\dot{q}_i(d)]} \tag{5.59b}$$

where

$$\begin{aligned}q_r(d)\dot{q}_i(d)-\dot{q}_r(d)q_i(d)= w^2(0)\{&H_{fr}(d)\dot{H}_{fi}(d)-\dot{H}_{fr}(d)H_{fi}(d)\\&+(a^2+b^2)[H_{ar}(d)\dot{H}_{ai}(d)-\dot{H}_{ar}(d)H_{ai}(d)]\\&+b[\dot{H}_{fr}(d)H_{ar}(d)+\dot{H}_{fi}(d)H_{ai}(d)\\&-H_{fr}(d)\dot{H}_{ar}(d)-H_{fi}(d)\dot{H}_{ai}(d)]\\&+a[H_{fr}(d)\dot{H}_{ai}(d)+\dot{H}_{fi}(d)H_{ar}(d)\\&-\dot{H}_{fr}(d)H_{ai}(d)-H_{fi}(d)\dot{H}_{ar}(d)]\}\end{aligned} \tag{5.60a}$$

$$\begin{aligned}q_r(d)\dot{q}_r(d)+q_i(d)\dot{q}_i(d)=&[1-c^2][H_{fr}(d)\dot{H}_{fr}(d)+H_{fi}(d)\dot{H}_{fi}(d)]\\&+[a(1-c^2)+2cb][H_{fr}(d)\dot{H}_{ar}(d)+\dot{H}_{fr}(d)H_{ar}(d)\\&-\dot{H}_{fi}(d)H_{ai}(d)-H_{fi}(d)\dot{H}_{ai}(d)]\\&+[(1-c^2)(a^2-b^2)+4abc][H_{ar}(d)\dot{H}_{ar}(d)-H_{ai}(d)\dot{H}_{ai}(d)]\\&+2[b(1-c^2)-2ac][H_{fr}(d)\dot{H}_{ai}(d)+H_{fi}(d)\dot{H}_{ar}(d)]\\&-2c[H_{fr}(d)\dot{H}_{fi}(d)+\dot{H}_{fr}(d)H_{fi}(d)]\\&+2[ab(1-c^2)-c(a^2-b^2)][H_{ar}(d)\dot{H}_{ai}(d)+\dot{H}_{ar}(d)H_{ai}(d)]\end{aligned} \tag{5.60b}$$

Note that in the case of active media, the z–invariant condition does not hold when the refractive index is a complex function [5.2]. This fact establishes an important difference from the passive case, in which the beam half–width is expressed as the modulus of the complex ray, and the wavefront is perpendicular to the beam profile. For comparison, see Eqs. (4.26–27), (4.35), and (4.37). In other words, we find that the Gaussian beam is not z–invariant when considering propagation through active media [5.26].

5.6 Transformation of a Gaussian Beam into a Uniform Beam

Based on the results of the preceding sections, we can study how to obtain uniform illumination from a Gaussian beam by an active GRIN medium. To achieve this transformation, we express Eqs. (5.59) as

$$w^2(d)=\frac{2\lambda|q(d)|^2}{\pi\left\{n_{0i}\left.\dfrac{d|q(z)|^2}{dz}\right|_{z=d}+2n_{0r}[q_r(d)\dot{q}_i(d)-\dot{q}_r(d)q_i(d)]\right\}} \tag{5.61a}$$

$$R(d)=\frac{2|q(d)|^2}{n_{0r}\left.\frac{d|q(z)|^2}{dz}\right|_{z=d}+2n_{0i}[\dot{q}_r(d)q_i(d)-q_r(d)\dot{q}_i(d)]} \tag{5.61b}$$

From Eq. (5.61a) it follows that, depending on the sign of the denominator, w^2 can be either positive or negative at the output face because the Gaussian beam is not z–invariant. If w^2 is positive, a Gaussian beam is obtained whose irradiance is a maximum at the axis (power concentration), and decreases with the transverse direction. If w^2 is negative, a reverse behavior occurs (power spread). Likewise, if the denominator is equal to zero, $w\to\infty$, and the condition for transforming a Gaussian beam into a uniform one is obtained. Therefore, this condition can be written as:

$$n_{0i}\left.\frac{d|q(z)|^2}{dz}\right|_{z=d}=-2n_{0r}[q_r(d)\dot{q}_i(d)-\dot{q}_r(d)q_i(d)] \tag{5.62}$$

The half–width of the input beam required to obtain uniform transformation is given by

$$w^2(0)=-\frac{n_{0r}[q_r(d)\dot{q}_i(d)-\dot{q}_r(d)q_i(d)]}{n_{0i}|G(d)||\dot{G}(d)|} \tag{5.63}$$

where Eq. (5.57) has been used.

Equation (5.63) indicates the relation between the half–width of the input beam and the parameters of the active medium for which a uniform spherical beam at the output face is obtained, and whose radius of curvature is expressed by

$$R(d)=\frac{2n_{0r}|q(d)|^2}{|n_0|^2\left.\frac{d|q(z)|^2}{dz}\right|_{z=d}}=\frac{n_{0r}|q(d)|}{|n_0|^2|\dot{q}(d)|} \tag{5.64}$$

On the other hand, from Eq. (5.64) two particular cases can be considered because the radius of curvature goes to infinity for $|\dot{q}(d)|=0$, and vanishes for $|q(d)|=0$. The first case provides the transformation of the input beam into a uniform plane beam at the output face of the active medium, and the second one the transformation into a point source. We can see that both cases are obtained as the position or the slope of the complex axial ray is cancelled at the output face [5.27].

Furthermore, setting the denominator of Eq. (5.61b) equal to zero gives the condition for transforming the input beam into a plane Gaussian beam at the output face of the active medium, that is

$$n_{0r}\left.\frac{d|q(z)|^2}{dz}\right|_{z=d}=-2n_{0i}[\dot{q}_r(d)q_i(d)-q_r(d)\dot{q}_i(d)] \tag{5.65}$$

or

$$w^2(0) = -\frac{n_{0i}[\dot{q}_r(d)q_i(d) - q_r(d)\dot{q}_i(d)]}{n_{0r}|G(d)||\dot{G}(d)|} \tag{5.66}$$

where Eq. (5.57) has been used.

Finally, when Eqs. (5.61a) and (5.65) are taken into account, the expression for the output beam waist can be written as

$$w_0^2(d) = \frac{2n_{0i}\lambda|q(d)|^2}{\pi|n_0|^2 \left.\frac{d|q(z)|^2}{dz}\right|_{z=d}} \tag{5.67}$$

Thus, Eq. (5.66) represents the half–width of the input beam needed to obtain a plane Gaussian beam of waist $w_0(d)$ at the output of the active GRIN medium. These results show the important role played by the active GRIN media in laser cavities and waveguides.

6 Planar GRIN Media with Hyperbolic Secant Refractive Index Profile

6.1 Introduction

Light propagation in GRIN media with a wide variety of gradient profiles has been analyzed with emphasis on focusing and collimation properties of importance for integrated optics, fiber optics, micro–optics, and optical sensing [6.1]. In designing GRIN structures, one goal is to select an index profile that provides focusing and collimation. The simplest GRIN medium with this capability is the parabolic one, where the equation governing light propagation can be solved analytically in the paraxial region [6.2], as discussed in Chap. 1. There is another GRIN medium with focusing and collimation properties. It is characterized by a hyperbolic secant profile (hereafter referred to as HS) that presents the following advantages in comparison with the parabolic one [6.2–6.5]. First, it is a more general profile, because the parabolic function can be considered as the first–order approximation of the Taylor expansion series of the hyperbolic secant function. Second, the equation governing light propagation can be solved analytically without carrying out any type of approximation. Third, the medium is free of aberrations for meridional rays (rays propagating in planes such that they include the optical axis). Fabrication of GRIN media with HS profiles has already been demonstrated by a microcontrolled dip–coating procedure from colloidal solutions [6.6]. Other methods include field–assisted ion exchange [6.7] and vapor deposition techniques [6.8].

In Chap. 6 we study light propagation by ray and modal theories in GRIN materials with HS profiles, as well as focusing and collimation properties of interest in designing devices for manipulating and processing optical signals.

6.2 Ray Equation and ABCD Law

We begin by introducing a GRIN medium whose transverse refractive index profile along the x–axis is given by

$$n^2(x) = n_s^2 + (n_0^2 - n_s^2)\operatorname{sech}^2(\alpha x) \tag{6.1}$$

representing a GRIN planar waveguide with HS profile in order to study light propagation. In Eq. (6.1), n_s is the homogeneous substrate's refractive index, n_0 is the index along the z–axis, α the gradient parameter, and α^{-1} is the half–width of the profile, that is, the distance in transverse direction at which the profile decays to 0.42 of its maximum value (Fig. 6.1).

The ray equation for describing light propagation through this kind of waveguide is obtained from the expression of a differential arc length along a ray joining any two points of the medium from Eq. (1.41), as shown in Fig. 6.2.

$$ds = \sqrt{1+\dot{x}^2}\,dz \tag{6.2}$$

Moreover, in a transverse gradient the third optical direction cosine l is invariant along any ray within the medium, and from Eq. (1.60) is given by

$$l = l_0 = n\frac{dz}{ds} = \frac{n}{(1+\dot{x}^2)^{1/2}} = \text{constant} \tag{6.3}$$

where Eq. (6.2) has been used.

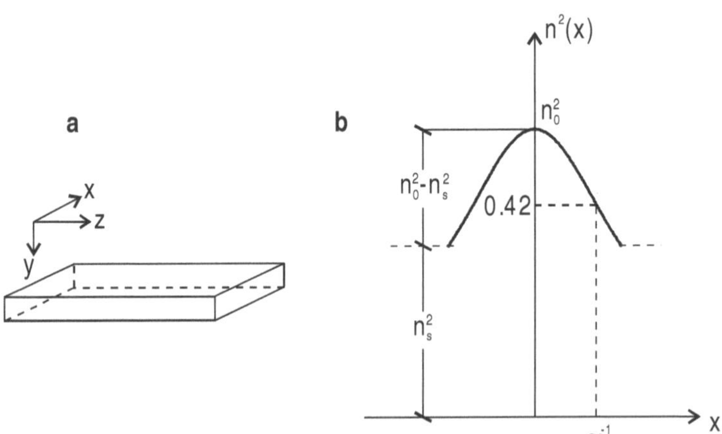

Fig. 6.1. **a** Planar waveguide and **b** hyperbolic secant refractive index profile.

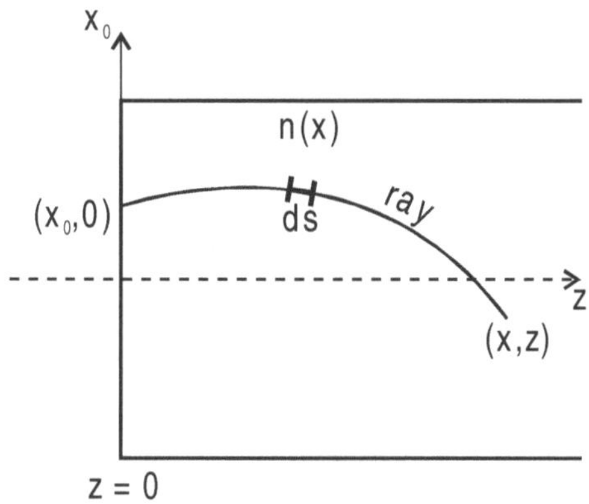

Fig. 6.2. Ray propagation in a GRIN planar waveguide with HS profile.

Equation (6.3) can be rewritten as

$$dz = \frac{l_0}{\sqrt{n^2 - l_0^2}} dx \tag{6.4}$$

Substitution of Eq. (6.1) into (6.4) provides

$$dz = \frac{\cosh(\alpha x)}{\gamma\sqrt{A^2 - \sinh^2(\alpha x)}} dx \tag{6.5}$$

where

$$\gamma = \frac{\sqrt{l_0^2 - n_s^2}}{l_0} \tag{6.6}$$

and

$$A^2 = \frac{n_0^2 - l_0^2}{l_0^2 - n_s^2} \tag{6.7}$$

From Eq. (6.4) it follows that the ray trajectory between any two points located at z = 0 and z > 0 is expressed by

$$z = \frac{1}{\gamma}\int_{x_0}^{x} \frac{\cosh(\alpha x)}{\sqrt{A^2 - \sinh^2(\alpha x)}} dx \tag{6.8}$$

To find the ray trajectory, we use a transformation of the Cartesian coordinate x into a hyperbolic coordinate

$$u = \sinh(\alpha x) \tag{6.9}$$

Under this transformation, substituting Eq. (6.9) into (6.8) and performing the integration with respect to u, the ray trajectory in the waveguide becomes

$$z = \frac{1}{\gamma\alpha}\left[\sin^{-1}\left(\frac{u}{A}\right) - \sin^{-1}\left(\frac{u_0}{A}\right)\right] \tag{6.10}$$

where u_0 is the value of u at $z = 0$, that is, $u_0 = \sinh(\alpha x_0)$.

Equation (6.10) can be rewritten as

$$u(z) = A\sin\tau \tag{6.11}$$

where

$$\tau = \sin^{-1}\left(\frac{u_0}{A}\right) + \gamma\alpha z \tag{6.12}$$

Here, u indicates the ray position at each point within the medium, and $\dot{u}$ represents the ray slope in the new coordinate system u–z, which is given by

$$\dot{u}(z) = \alpha\dot{x}\cosh(\alpha x) = \alpha\gamma A\cos\tau \tag{6.13}$$

where Eqs. (6.9) and (6.11–6.12) have been used.

The ray equation for position and slope given by Eqs. (6.11) and (6.13) is also written as

$$u(z) = A\left\{\frac{u_0}{A}\cos(\gamma\alpha z) + \cos\left[\sin^{-1}\left(\frac{u_0}{A}\right)\right]\sin(\gamma\alpha z)\right\} \tag{6.14a}$$

$$\dot{u}(z) = \alpha\gamma A\left\{\cos\left[\sin^{-1}\left(\frac{u_0}{A}\right)\right]\cos(\gamma\alpha z) - \frac{u_0}{A}\sin(\gamma\alpha z)\right\} \tag{6.14b}$$

but

$$\cos\left[\sin^{-1}\left(\frac{u_0}{A}\right)\right] = \cos\tau_0 = \frac{\dot{u}_0}{\alpha\gamma A} \tag{6.15}$$

where τ_0 and $\dot{u}_0$ are the values of τ and $\dot{u}$ at $z = 0$, respectively.

Then, Eq. (6.14) becomes

$$u(z) = u_0\cos(\alpha\gamma z) + \dot{u}_0\frac{\sin(\alpha\gamma z)}{\alpha\gamma} \tag{6.16a}$$

$$\dot{u}(z) = -u_0\alpha\gamma\sin(\alpha\gamma z) + \dot{u}_0\cos(\alpha\gamma z) \tag{6.16b}$$

and if we introduce the position and slope of the axial and field rays

$$H_a(z) = \frac{\sin(\alpha\gamma z)}{\alpha\gamma}; \quad \dot{H}_a(z) = \cos(\alpha\gamma z) \tag{6.17a}$$

$$H_f(z) = \cos(\alpha\gamma z); \quad \dot{H}_f(z) = -\alpha\gamma\sin(\alpha\gamma z) \tag{6.17b}$$

Equation (6.16) reduces to

$$u(z) = u_0 H_f(z) + \dot{u}_0 H_a(z) \tag{6.18a}$$

$$\dot{u}(z) = u_0 \dot{H}_f(z) + \dot{u}_0 \dot{H}_a(z) \tag{6.18b}$$

The initial conditions for these rays are

$$H_a(0) = \dot{H}_f(0) = 0 \tag{6.19a}$$

$$\dot{H}_a(0) = H_f(0) = 1 \tag{6.19b}$$

and Lagrange's invariant says

$$\dot{H}_a(z) H_f(z) - H_a(z) \dot{H}_f(z) = 1 \tag{6.20}$$

Equation (6.18) has the same form as the paraxial ray equation (1.77) for a parabolic GRIN medium, but in this case the ray equation was obtained, in the new coordinate system u–z, without any approximation. This confirms that a planar medium with HS profile is free of aberrations.

Taking into account Eqs. (6.18), (6.13), and (6.11), we can determine the real ray position x and the real ray slope at z, that is

$$x(z) = \frac{1}{\alpha}\sinh^{-1}[u(z)] = \frac{1}{\alpha}\sinh^{-1}[u_0 H_f(z) + \dot{u}_0 H_a(z)] \tag{6.21a}$$

$$\dot{x}(z) = \frac{\dot{u}(z)}{\alpha\cosh[\alpha x(z)]} = \frac{u_0 \dot{H}_f(z) + \dot{u}_0 \dot{H}_a(z)}{\alpha\cosh\{\sinh^{-1}[u_0 H_f(z) + \dot{u}_0 H_a(z)]\}} \tag{6.21b}$$

Likewise, the linear algebraic nature of Eq. (6.18) allows us to use 2×2 matrices in the system u–z for describing light propagation in the planar waveguide. The resulting matrix equation is of the form

$$\begin{pmatrix} u(z) \\ \dot{u}(z) \end{pmatrix} = \begin{pmatrix} H_f(z) & H_a(z) \\ \dot{H}_f(z) & \dot{H}_a(z) \end{pmatrix} \begin{pmatrix} u_0 \\ \dot{u}_0 \end{pmatrix} \tag{6.22}$$

Equation (6.22) is analogous to the ABCD law (see Eqs. (1.89)) used in geometrical optics for analyzing light propagation through any system within the framework of the paraxial approximation.

6.3 Focusing and Collimation Properties

From the oscillatory nature of the basic rays H_a, H_f it follows that any ray leaving the input of a planar waveguide with HS profile describes a sinusoidal path through it. This fact permits us to study focusing and collimation properties in this kind of planar waveguide. Likewise, as mentioned in Chap. 2, there is a method for determining the image and transform conditions in GRIN media based on the evaluation of the zeros of functions H_a and H_f. This method can be applied for obtaining the focusing and collimation conditions for a planar waveguide with HS profile when it is illuminated by a point source at the input [6.9].

On the one hand, for a point source at the input z = 0 of the waveguide, the focusing condition is given by

$$H_a(z_m) = \dot{H}_f(z_m) = 0 \tag{6.23}$$

where m is an integer, and z_m are the waveguide lengths satisfying

$$z_m = \frac{m\pi}{\alpha\gamma} \tag{6.24}$$

At this condition, the ray position and ray slope are expressed as

$$x(z_m) = (-1)^m x_0 \tag{6.25a}$$

$$\dot{x}(z_m) = (-1)^m \dot{x}_0 \tag{6.25b}$$

Thus, the ray position at z_m remains constant, and it is independent of the ray slope at the input. In other words, planar waveguides with HS profiles have perfect equalization at z_m for all rays. Cutting planar waveguides at lengths z_m, we

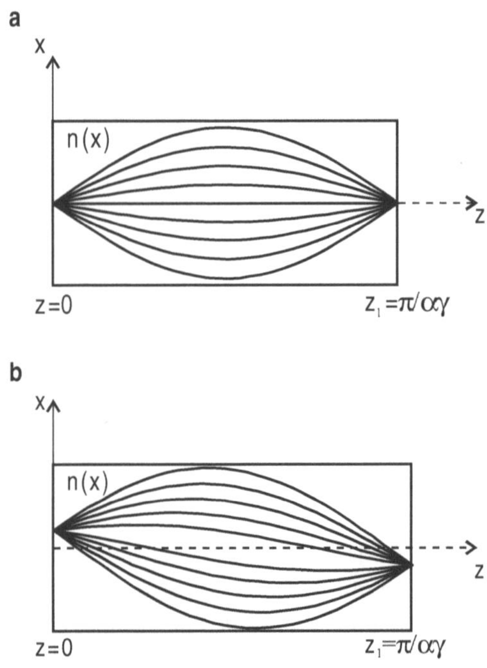

Fig. 6.3. On–axis **a** and off–axis **b** image formation: light focuser.

can obtain stigmatic focusing for a source located at the input, and the waveguides may be applied as light focusing devices (Fig. 6.3).

On the other hand, the collimation condition for a source located at z = 0 is given by

$$H_f(z_p) = \dot{H}_a(z_p) = 0 \tag{6.26}$$

where p is an integer, and z_p are the waveguide lengths satisfying

$$z_p = \frac{(2p+1)\pi}{2\alpha\gamma} \tag{6.27}$$

Ray position and ray slope are now expressed as

$$x(z_p) = \frac{(-1)^p}{\alpha} \sinh^{-1}\left[\frac{\dot{x}_0}{\gamma} \cosh(\alpha x_0)\right] \tag{6.28a}$$

$$\dot{x}(z_p) = (-1)^{p+1} \frac{\sinh(\alpha x_0)}{\gamma \cosh\left\{\sinh^{-1}\left[\frac{\dot{x}_0}{\gamma}\cosh(\alpha x_0)\right]\right\}} \tag{6.28b}$$

From Eq. (6.28) it follows that we obtain perfect collimation only if $x_0 = 0$ since these equations become

$$x(z_p) = \frac{(-1)^p}{\alpha} \sinh^{-1}\left[\frac{\dot{x}_0}{\gamma}\right] \tag{6.29a}$$

$$\dot{x}(z_p) = 0 \tag{6.29b}$$

Then, at z_p all the rays will be parallel to the z–axis for an on–axis source at the input of planar waveguide. The waveguide at lengths z_p works like a light collimating device (Fig. 6.4).

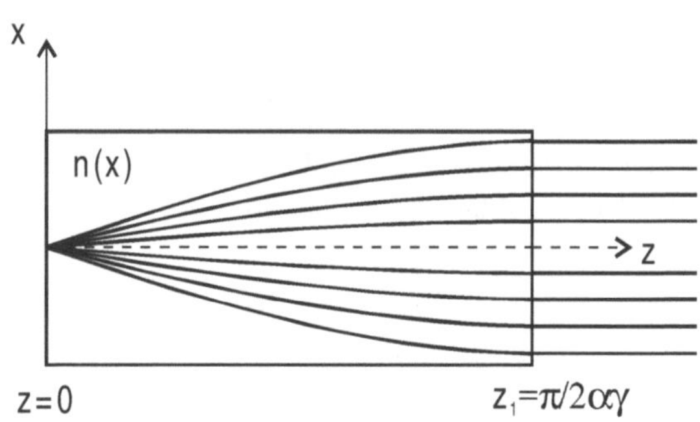

Fig. 6.4. Collimation for an on–axis source: light collimator.

Note that, from the principle of reversibility, a collimated beam propagating along the z–axis at $z = 0$ will be focused at z_p just as an on–axis source at $z = 0$ will produce a parallel beam of rays at z_p. So $H_f(z_p) = 0$ is also the focusing condition when a collimated beam propagating along the z–axis impinges on the input of the waveguide.

Optical properties such as focusing and collimation are of great importance in designing optical devices for integrated optics or micro–optics , such as fiber–fiber, source–fiber, or fiber–detector connectors, directional couplers, and so on. Using planar media with HS profiles having these properties, light deflector and beam–size controller devices can be designed [6.9–6.11].

We consider the butt–joining coupling of two planar waveguides of lengths d_1 and d_2 whose refractive index profiles $n_1(x)$ and $n_2(x)$, respectively, are given by functions such as Eq. (6.1). Figure 6.5 shows the geometry of the problem and the coordinate system. Assuming that the z–axis refractive indices of both waveguides are different ($n_{01} \neq n_{02}$) and that the same occurs for the gradient parameters ($\alpha_1 \neq \alpha_2$), the input position and slope in the second waveguide, in hyperbolic space, expressed as a function of the output position and slope in the first waveguide can be written as [6.10]

$$u_{02}(d_1) = \sinh\left[\frac{\alpha_2}{\alpha_1}\sinh^{-1}[u_1(d_1)]\right] \tag{6.30a}$$

$$\dot{u}_{02}(d_1) = \frac{l_{01}\alpha_2 \cosh[\alpha_2 x_1(d_1)]}{l_{02}\alpha_1 \cosh[\alpha_1 x_1(d_1)]} \tag{6.30b}$$

where Snell's law for the interface between waveguides

$$l_{01}\dot{x}_1(d_1) = l_{02}\dot{x}_{02}(d_1) \tag{6.31}$$

and Eqs. (6.9), (6.13), and (6.21) have been used, and l_{01} and l_{02} are the third optical direction cosines in both waveguides.

From Eqs. (6.22) and (6.30) it follows that the ray equation for the second waveguide is expressed, in matrix form, as

$$\begin{pmatrix} u_2(z) \\ \dot{u}_2(z) \end{pmatrix} = \begin{pmatrix} H_{f2}(z) & H_{a2}(z) \\ \dot{H}_{f2}(z) & \dot{H}_{a2}(z) \end{pmatrix} \begin{pmatrix} u_{02}(d_1) \\ \dot{u}_{02}(d_1) \end{pmatrix} \tag{6.32}$$

where subscript 2 denotes the second waveguide.

Assuming that the lengths of the waveguides are

$$d_1 = {(2p_1+1)\pi}/{2\alpha_1} \tag{6.33a}$$

$$d_2 = {(2p_2+1)\pi}/{2\alpha_2} \tag{6.33b}$$

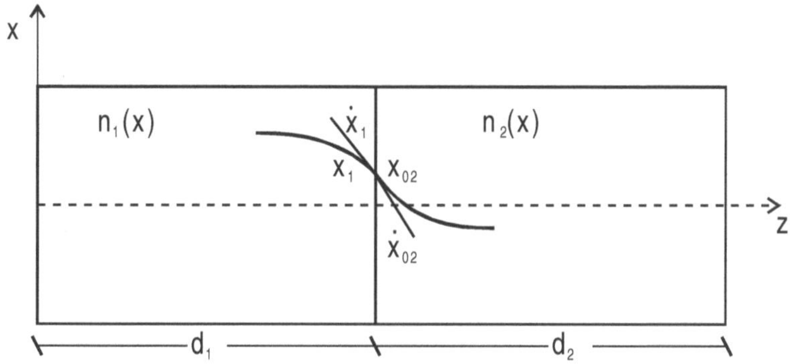

Fig. 6.5. Geometry and coordinate system for butt–joining coupling of two planar waveguides with HS profiles.

where p_1 and p_2 are integers, and that the coupling system is illuminated by a collimated beam of rays propagating parallel to the z–axis, Eq. (6.32) becomes

$$\begin{pmatrix} u_2(d_1+d_2) \\ \dot{u}_2(d_1+d_2) \end{pmatrix} = \begin{pmatrix} 0 & \dfrac{(-1)^{p_2}}{\alpha_2\gamma_2} \\ (-1)^{p_1+p_2}\alpha_2\gamma_2 & 0 \end{pmatrix} \begin{pmatrix} 0 \\ (-1)^{p_1+1} l_{01}\gamma_1\alpha_2 u_{01} \Big/ l_{02} \end{pmatrix} \tag{6.34}$$

where the initial conditions

$$u_1(z=0) = u_{01}\,; \quad \dot{u}_1(z=0) = \dot{u}_{01} = 0 \tag{6.35}$$

the interface conditions

$$u_1(d_1) = u_{02}(d_1) = 0; \quad \dot{u}_1(d_1) = (-1)^{p_1+1}\alpha_1\gamma_1 u_{01} \tag{6.36}$$

and Eqs. (6.17) and (6.30) have been used.

Note that Eq. (6.26) is the focusing condition for the first waveguide of length d_1 and is the collimation condition for the second waveguide of length d_2.

From Eq. (6.34) it follows that at the output of the coupling system we have a collimated beam propagating in the same direction as the input beam, that is

$$\dot{u}_2(d_1+d_2) = 0 \tag{6.37}$$

whose ray positions are given by

$$u_2(d_1+d_2) = (-1)^{p_1+p_2+1}\frac{l_{01}\gamma_1}{l_{02}\gamma_2}u_{01} \tag{6.38}$$

as shown in Fig. 6.6.

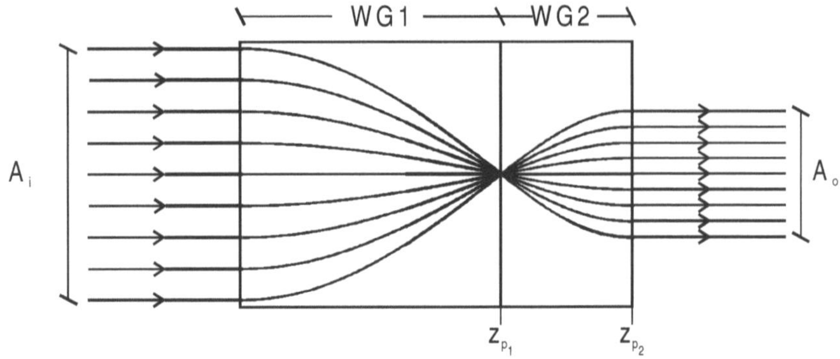

Fig. 6.6. Beam–size contraction by butt–joining of planar waveguides with HS profiles.

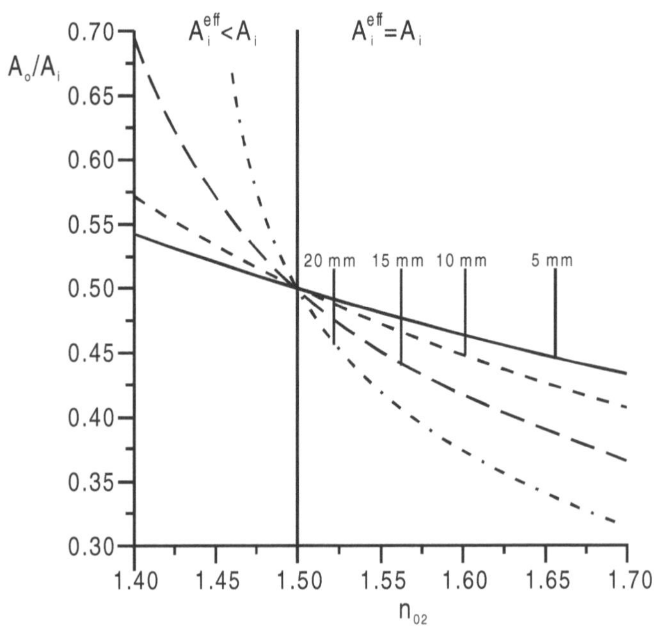

Fig. 6.7. Variation of the input and output beam aperture relationship versus n_{02} for different values of the input marginal ray. Calculations have been made for $n_{01} = 1.5$, $\alpha_1 = 0.1\,\mathrm{mm}^{-1}$, and $\alpha_2 = 2\alpha_1$.

However, the apertures of the input and output beams are not equal, and the relationship between them can be obtained from the input marginal ray, denoted by u_{0m}. This relationship is given by

$$\frac{A_o}{A_i} = \frac{1}{\alpha_2 x_{0m}} \sinh^{-1}\left[\frac{l_{01}\gamma_1}{l_{02}\gamma_2} u_{0m}\right] \tag{6.39}$$

where A_o and A_i are the input and output apertures, and x_{0m} is the real position of the marginal ray at the input of the coupling system.

Figure 6.7 depicts variation of the aperture relationship versus n_{02} for different height values of the input marginal ray and $n_{01} = 1.5$. Note that the cutoff point of the curves represents the critical angle at the interface, and that we can contract the beam size by choosing the appropriate values of waveguide parameters. Likewise, from the principle of reversibility of rays, we can also expand the beam size. Therefore this butt–coupling system works like a beam–size controller device.

Furthermore, it is easy to prove that, depending on the coupled waveguides' lengths, it is also possible to design focusers and collimators. Figure 6.8 is a block diagram summarizing the devices obtained for different lengths of coupled waveguides and different kinds of illumination. For instance, a device made by coupling two planar waveguides with HS profiles and $\pi/2\alpha_1$, π/α_2 lengths, respectively, can operate as a collimator for point source illumination or as a focuser for collimated beam illumination.

Now we will analyze light propagation through a planar waveguide with HS profile illuminated by a tilted plane beam. We assume that a tilted beam of parallel rays impinges on the input of the waveguide, making an angle β with the z–axis. In Fig. 6.9a we show the ray–tracing for this tilted beam through the waveguide. The ray–tracing is periodic, and we can see that the trajectories of the rays in two consecutive periods are symmetrical with respect to the z–axis. In Fig. 6.9b we can observe that aberrations lead to a focal spot and not to a stigmatic point. This spot is a minimum at planes that satisfy the collimation condition, that is, at planes z_p given by Eq. (6.26).

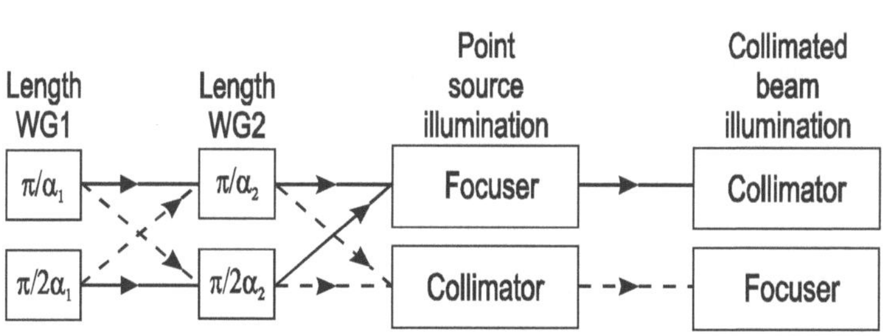

Fig. 6.8. Block diagram for obtaining a focuser and a collimator by two coupled planar waveguides with HS profile.

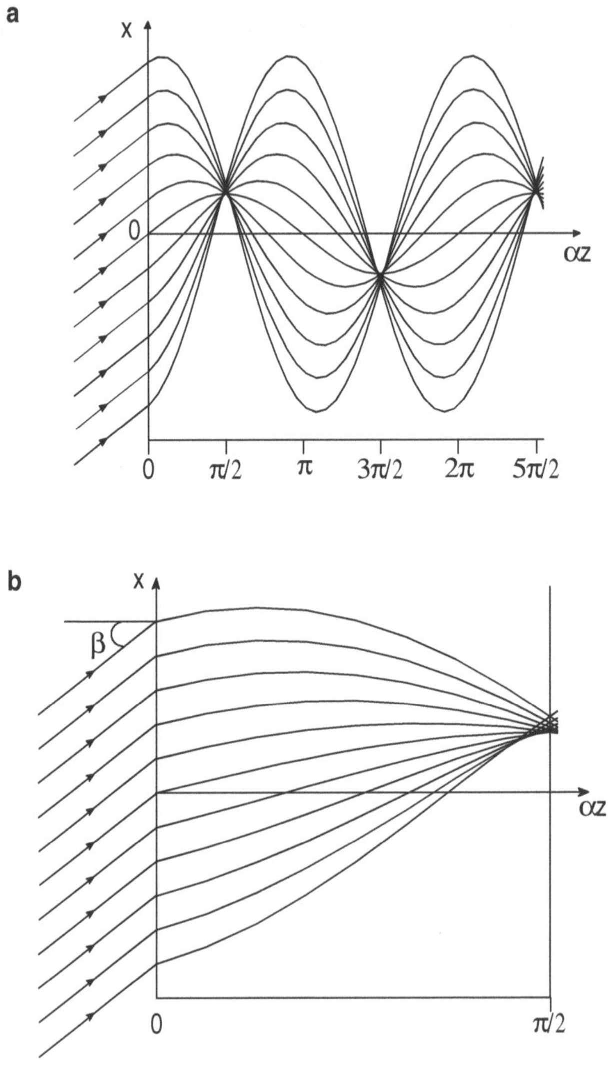

Fig. 6.9. **a** Propagation of a tilted plane beam through a planar HS waveguide; **b** Focal spot produced by a tilted plane beam on the plane $z = \pi/2\alpha\gamma$ (p=0).

On the other hand, we can analyze the behavior of the light at planes z_m for tilted illumination. In this case, the initial conditions for any ray are given by

$$x(z=0)=x_0 ; \quad \dot{x}(z=0)=\dot{x}_0 \tan\gamma \tag{6.40}$$

where γ is the refraction angle given by Snell's law, which can be written as

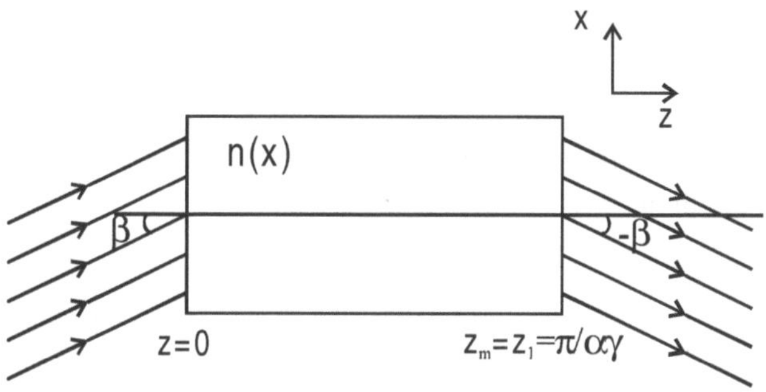

Fig. 6.10. Light deflection by a planar HS waveguide.

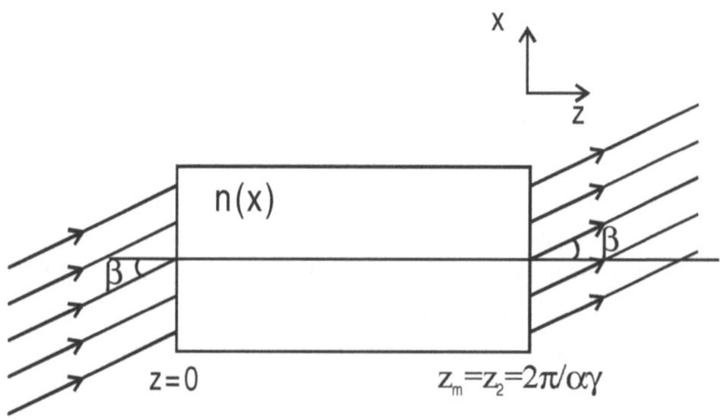

Fig. 6.11. Light shifting by a planar HS waveguide.

$$\sin\beta = l_0 \dot{x}_0 \tag{6.41}$$

provided that the waveguide is surrounded by vacuum or air.

Therefore, at planes z_m the real position and slope of the rays are expressed as

$$x(z_m) = (-1)^m x_0 \tag{6.42a}$$

$$\dot{x}(z_m) = (-1)^m \frac{\sin\beta}{l_0} \tag{6.42b}$$

From Eq. (6.42) it follows that at planes z_m the position and slope of the rays depend on the mth–order, that is, on the waveguide length, and that at any plane z_m the slope of the ray is a constant.

Thus, if we cut the planar waveguide with HS profile at planes z_m with m an odd integer (m = 2q+1), the exit beam will be a plane beam tilted at an angle –β as shown in Fig. 6.10. In other words, cutting the waveguide at these planes, we can obtain a light deflector of angle 2β. An application for this device is the substitution of a mirror in an optical setup, and, in general, it may be used for changing the light propagation direction.

Likewise, if we cut the waveguide at planes z_m with m an even integer (m=2q), the exit beam will have the same direction as the input one. Therefore, the waveguide only moves the input beam as shown in Fig. 6.11 and works like a light shifter.

In short, the focuser, the collimator, the beam–size controller, the light deflector, and the light shifter that have been presented are five important options for designers of integrated optics to take into account.

6.4 Numerical Aperture: On–Axis and Off–Axis Coupling

As commented in Sect. 6.3, a planar waveguide with HS profile can carry out typical functions such as collimation and focusing, which are of great importance in integrated optics. Taking advantage of these inherent functions, some devices have been designed. Likewise, the on–axis and off–axis imaging properties of this kind of waveguide can be used for designing on–axis and off–axis couplers.

For on–axis butt–joining connection by planar waveguides with HS profiles, the main optical concept is the numerical aperture (NA). Now we consider a waveguide of semiaperture a. Figure 6.12 shows the geometry of the problem and the trajectory of the marginal ray for an on–axis source located at a certain distance from the input in order to evaluate the NA. The input NA related to the object space is given by

$$NA_{in}(x_{0m}) = \sin\theta_{0m} \tag{6.43}$$

provided that the waveguide is immersed in vacuum or air.

From Snell's law at the input of the waveguide, Eq. (6.43) becomes

$$NA_{in}(x_{0m}) = l(x_{0m})\dot{x}_{0m} \tag{6.44}$$

where

$$l(x_{0m}) = n(x_{0m})(1+\dot{x}_{0m}^2)^{-1/2} \tag{6.45}$$

and $\dot{x}_{0m}$ is the input slope of the marginal ray.

After the marginal ray has propagated a length d in the waveguide, the ray achieves a maximum value where the boundary conditions for light confinement due to the waveguide semiaperture

$$x_i = a \Rightarrow u_i(d) = \sinh(\alpha a) = u_{0m}H_f(d) + \dot{u}_{0m}H_a(d) \tag{6.46a}$$

$$\dot{x}_i = 0 \Rightarrow \dot{u}_i(d) = 0 = u_{0m}\dot{H}_f(d) + \dot{u}_{0m}\dot{H}_a(d) \tag{6.46b}$$

are satisfied.

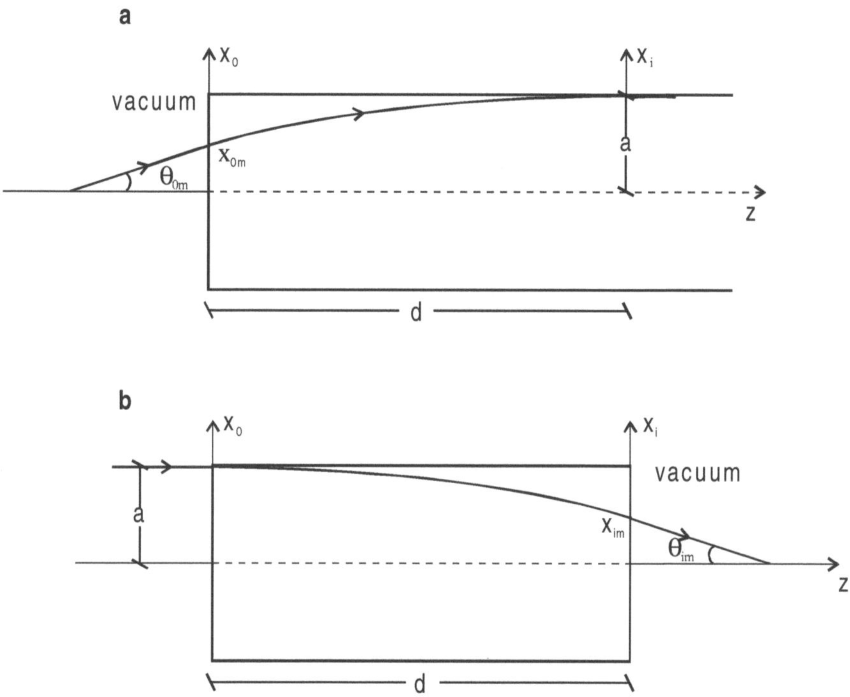

Fig. 6.12. a Input and **b** output NAs for a planar waveguide with HS profile. The input and output NAs are defined in the object and image spaces, respectively.

From Eq. (6.46b) it follows that

$$u_{0m} = -\frac{\dot{H}_a(d)}{\dot{H}_f(d)}\dot{u}_{0m} \tag{6.47}$$

Substituting Eq. (6.46a) and taking into account Eq. (6.20), we have

$$u_i(d) = -\frac{\dot{u}_{0m}}{\dot{H}_f(d)} = -\frac{\alpha\dot{x}_{0m}\cosh(\alpha x_{0m})}{\dot{H}_f(d)} \tag{6.48}$$

Equation (6.48) can be rewritten as

$$u_i(d) = \frac{u_{0m}}{\dot{H}_a(d)} = \frac{\sinh(\alpha x_{0m})}{\dot{H}_a(d)} \tag{6.49}$$

where Eq. (6.47) has been used.

Position and slope of the input marginal ray can be obtained from Eqs. (6.48–6.49) to give

$$x_{0m} = \frac{1}{\alpha}\sinh\{\sin(\alpha\gamma d)\sinh(\alpha a)\} \tag{6.50a}$$

$$\dot{x}_{0m} = \gamma\sin(\alpha\gamma d)\sinh(\alpha a)\mathrm{sech}(\alpha x_{0m}) \tag{6.50b}$$

where Eqs. (6.17) have been used, and x_{0m} can also be regarded as the effective semiaperture of the planar waveguide at the input.

Inserting Eq. (6.50b) into Eq. (6.44), we obtain

$$NA_{in}(x_{0m}) = \frac{\gamma n(x_{0m})\mathrm{sech}(\alpha x_{0m})\sin(\alpha\gamma d)\sinh(\alpha a)}{\sqrt{1+\gamma^2\mathrm{sech}^2(\alpha x_{0m})\sin^2(\alpha\gamma d)\sinh^2(\alpha a)}} \tag{6.51}$$

Equation (6.51) indicates that the NA depends on a set of parameters such as the refractive index profile, the semiaperture, and the length of the waveguide. These parameters play an important role in planar waveguides with HS profiles for coupling purposes.

For an on–axis source located at the input ($x_{0m} = 0$), the marginal ray achieves the maximum value at lengths for which the collimation condition given by Eq. (6.27) is fulfilled. In this case, the NA reduces to

$$NA_{in}(0) = (-1)^p \frac{\gamma n_0\sinh(\alpha a)}{\sqrt{1+\gamma^2\sinh^2(\alpha a)}} \tag{6.52}$$

This result, for instance, can be applied to the on–axis butt–joining coupling between a planar waveguide with HS profile and a single planar structure.

On the other hand, the NA related to the image space, that is, the output NA (Fig. 6.12b) is given by

$$NA_{out}(x_{im}) = \sin\theta_{im} = l(x_{im})\dot{x}_{im} \tag{6.53}$$

where

$$l(x_{im}) = n(x_{im})[1+\dot{x}^2_{im}]^{-1/2} \tag{6.54}$$

and $\dot{x}_{im}$ is the output slope of the marginal ray.

In this case, the output NAs for the on–axis images located at a certain distance from the output or located at the output coincide with Eqs. (6.51) and (6.52), respectively, because the marginal ray positions are equal at the input and output of the waveguide and l_0 is invariant along any ray. Therefore, for instance, an optical configuration in which a light beam coming from an on–axis source at the input is focused at the output, as shown in Fig. 6.13, can be applied to the design of an on–axis butt–joining connector for unitary NA conversion between two identical single planar structures by a waveguide with HS profile.

In the same way, for off–axis butt–joining connections the main optical concept is the ray transformation for position and slope that is a unitary transformation taking into account Eq. (6.25). Figure 6.13 also depicts an off–axis connection by a planar waveguide with HS profile.

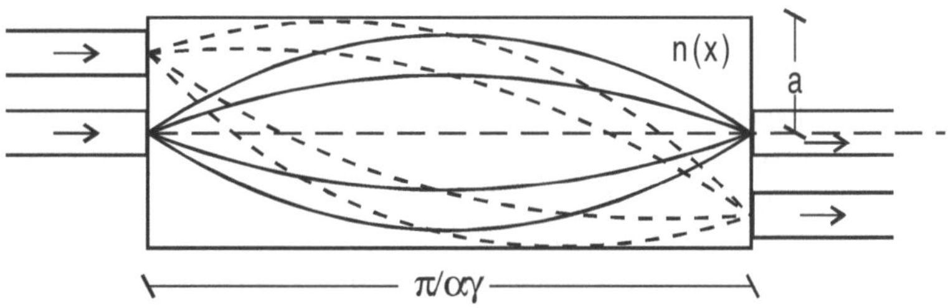

Fig. 6.13. On–axis (solid line) and off–axis (dashed line) coupling between planar structures by a planar waveguide with HS profile of length d=π/αγ. Waveguide works like unitary transformer for NAs, position, and slope of rays.

The optical configurations presented can be also applied to the design of other typical devices for integrated optics, such as beam divider using a half–mirror or filter, directional coupler, wavelength–division multiplexer, and so on, by two planar waveguides with HS profiles of quarter–pitch (p = 0) arranged in series.

Finally, these results can be extended to tapered planar waveguides characterized by a transverse hyperbolic secant refractive index modulated by an axial index along the z–axis [6.11–6.14]. In particular, a beam–size controller can be designed by only one waveguide. The device obtained has an important advantage in comparison with the nontapered device, since the total reflection on the interface between waveguides is avoided.

6.5 Mode Propagation

In the scalar approximation, the propagation of transverse electric (TE) modes in an inhomogeneous linear waveguide with one–dimensional transverse geometry for the HS refractive index profile is described by

$$\left(\frac{\partial^2}{\partial x^2}+\frac{\partial^2}{\partial z^2}\right)\psi(x,z)+k^2n^2(x)\psi(x,z)=0 \tag{6.55}$$

where $n^2(x)$ is given by Eq. (6.1), ψ(x,z) represents the field distribution in the waveguide, and k is the wavenumber in vacuum.

For a field distribution that takes the form of waves traveling in the z–direction with a propagation constant β, that is, for

$$\psi(x,z)=\phi(x)\,e^{i\beta z} \tag{6.56}$$

Eq. (6.55) becomes

$$\left[\frac{d}{dx^2}+k^2n^2(x)\right]\phi(x)=\beta^2\phi(x) \tag{6.57}$$

where $\phi(x)$ is the transverse amplitude of the light spatial distribution. The wave equation (6.56) has the same form as the one–dimensional Schrödinger equation of quantum mechanics in a hyperbolic potential distribution, taking into account Eq. (6.1). The wave equation reveals a close relationship between the transverse amplitudes and propagation constants of guided modes in the waveguide and the bound eigenstates and eigenvalues of the Schrödinger equation [6.15–6.19].

In order to obtain the transverse amplitude $\phi_v(x)$ of guided modes, substitution of Eq. (6.1) into Eq. (6.57) provides

$$\frac{d^2\phi_v(x)}{dx^2}+\alpha^2\{V^2\text{sech}^2(\alpha x)-W^2\}\phi_v(x)=0 \tag{6.58}$$

where V is

$$V=\frac{k}{\alpha}\sqrt{n_0^2-n_s^2} \tag{6.59}$$

the normalized width of the profile such that

$$V^2=U^2+W^2=S(S+1) \tag{6.60}$$

and

$$U=\frac{1}{\alpha}\left(k^2n_0^2-\beta_v^2\right)^{1/2};\ W=\frac{1}{\alpha}(\beta_v^2-k^2n_s^2)^{1/2} \tag{6.61}$$

are the mode parameters for the waveguide and the substrate, respectively, $\nu=0,1,2,...$ the mode order, and

$$S=\frac{1}{2}\left[\sqrt{1+4V^2}-1\right] \tag{6.62}$$

If we introduce the changes of field amplitude and variable

$$\phi_v=[1-\tanh^2(\alpha x)]f_v \tag{6.63}$$

$$v=\frac{1}{2}[1-\tanh(\alpha x)] \tag{6.64}$$

Equation (6.58) becomes the differential equation

$$v(1-v)\frac{d^2f_v}{dv}+(W+1)(1-2v)\frac{df_v}{dv}-(W-S)(W+S+1)f_v=0 \tag{6.65}$$

which is a hypergeometric equation [6.20].

Solution of Eq. (6.65) is given by

$$f_v(v)=F(W-S,W+S+1,W+1;v) \tag{6.66}$$

with the following condition

$$W-S=-\nu \tag{6.67}$$

F is the hypergeometric function, which for integer ν is a polynomial of degree ν

$$F(-\nu, W+S+1, W+1; v) = \frac{(1-v)^{\nu-S}\, v^{\nu-S}}{(W+1)(W+2)\cdots(W+\nu)} \frac{d^{\nu}}{dv^{\nu}}\left[v^{W+\nu}(1-v)^{W+\nu}\right] \tag{6.68}$$

Then, the amplitude of a ν–order guided mode can be written as

$$\phi_{\nu}(x) = \mathrm{sech}^{S-\nu}(\alpha x) F\left(-\nu, 2S-\nu, S-\nu+1; \frac{1}{2}[1-\tanh(\alpha x)]\right) \tag{6.69}$$

where Eqs. (6.63–6.64) and (6.66–6.67) have been used.

On the other hand, the propagation constant of a guided mode can be determined by inserting Eqs. (6.60–6.61) into (6.67) and is given by

$$\beta_{\nu}^2 = k^2 n_s^2 + \alpha^2 (S-\nu)^2 \tag{6.70}$$

and there is a discrete and finite set of propagation constants, defined in the interval

$$kn_s \le \beta_{\nu} \langle kn_0 \tag{6.71}$$

Figure 6.14 shows the variation of the propagation constant with the mode order. Only ten modes in the waveguide can be confined.

These results allow us to write the spatial distribution of a ν–order guided mode as

$$\psi_{\nu}(x,z) = \mathrm{sech}^{S-\nu}(\alpha x) \cdot F\left(-\nu, 2S-\nu, S-\nu+1; \frac{1}{2}[1-\tanh(\alpha x)]\right) \exp\left\{ ikn_s z \sqrt{1 + \frac{\alpha^2 (S-\nu)^2}{k^2 n_s^2}} \right\} \tag{6.72}$$

The field distributions for the two lower–order modes are given by

$$\phi_0(x,z) = \mathrm{sech}^{S}(\alpha x) \exp\left\{ ikn_s z \sqrt{1 + \frac{\alpha^2 S^2}{k^2 n_s^2}} \right\}; \quad \nu = 0 \tag{6.73}$$

$$\phi_1(x,z) = \mathrm{sech}^{S}(\alpha x) \sinh(\alpha x) \exp\left\{ ikn_s z \sqrt{1 + \frac{\alpha^2 (S-1)^2}{k^2 n_s^2}} \right\}; \quad \nu = 1 \tag{6.74}$$

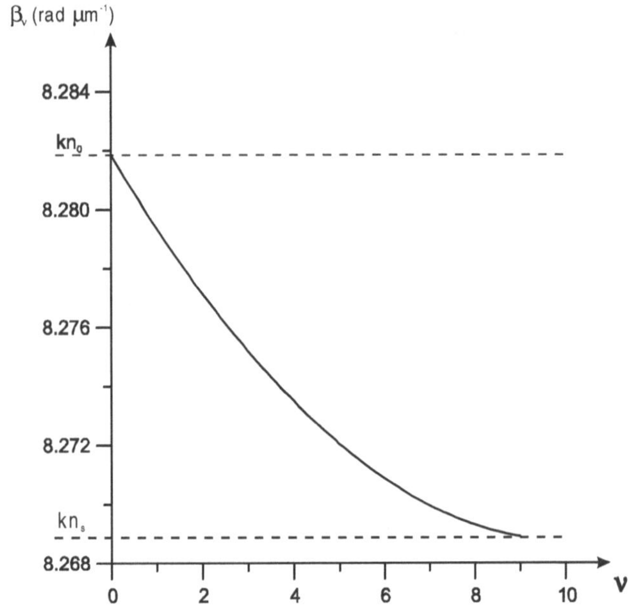

Fig. 6.14. Variation of the propagation constant β with the order of the guided mode for $n_0 = 1.45$, $n_s = 1.4476$, $\lambda = 1.1\,\mu m$, $\alpha^{-1} = 21.5\,\mu m$, and S = 9.

Likewise, the mode cutoff is defined by the lower limit $\beta_\nu = kn_s$ of bound–mode propagation constant values. So, the cutoff condition can be expressed in terms of the mode parameters and normalized width as

$$W = 0; \quad V = U \tag{6.75}$$

The cutoff value V_c for a multimode waveguide propagating up to ν–order mode can be obtained from Eqs. (6.75), (6.70), and (6.61) to give

$$V_c = \sqrt{(\nu+1)(\nu+2)} \tag{6.76}$$

and for this cutoff value, the half–width of the HS profile is

$$\alpha_c^{-1} = \frac{1}{k}\sqrt{\frac{(\nu+1)(\nu+2)}{n_0^2 - n_s^2}} \tag{6.77}$$

Then V governs the number of guided modes and when $V \leq \sqrt{2}$, that is, $S \leq 1$, all modes with the exception of the fundamental mode (ν = 0) are cut off.

Figure 6.15 depicts the amplitude distribution in trimode (a), bimode (b), and monomode (c) planar waveguides with HS profiles. Calculations have been made for $n_0 = 1.45$; $n_s = 1.4476$; $\lambda = 1.1\,\mu m$ and $\alpha = 0.22\,\mu m^{-1}$.

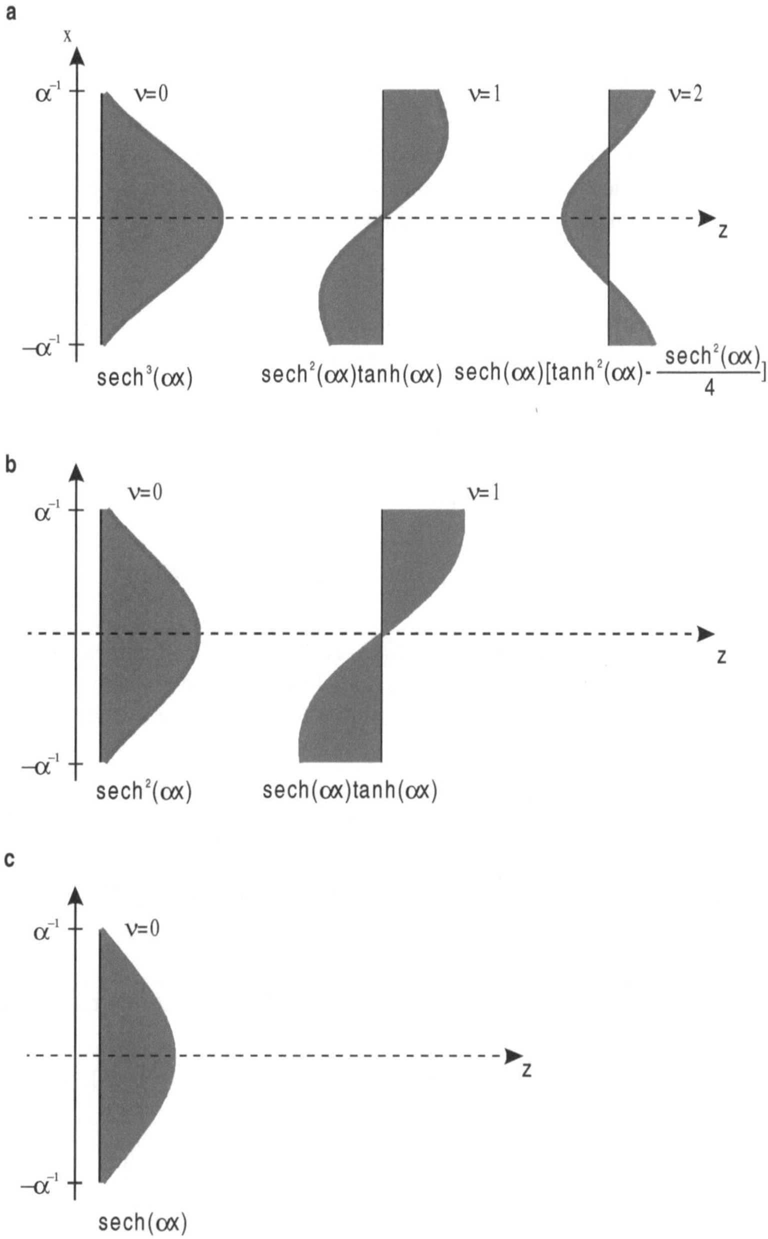

Fig. 6.15. Spatial distribution of **a** fundamental, first, and second order modes of a trimode waveguide for $S = 3$; **b** fundamental and first–order modes of a bimode waveguide for $S = 2$; **c** fundamental mode of a monomode waveguide for $S = 1$.

6.6 The Kernel Function

The wave analysis of light propagation through a planar waveguide with HS profile can be made by the kernel function K associated with the elliptic wave equation (6.55) with the boundary condition

$$K(x, x_0; z) = \delta(x - x_0) \quad \text{as} \quad z \to 0 \tag{6.78}$$

Then, if ψ_0 represents the field distribution at the input z = 0, the complex amplitude distribution at an arbitrary value z > 0 is expressed by the integral equation

$$\psi(x, z) = \int_{\Re} \psi_0(x_0) K(x, x_0; z)\, dx_0 \tag{6.79}$$

where, from Eq. (1.22), K is given by

$$K(x, x_0; z) = F_0(x_0) \exp\left\{-\frac{1}{2l_0}\int_0^z \nabla_\perp^2 P dz'\right\} \exp\{ikP\} \tag{6.80}$$

where Eq. (6.3) has been used, $\nabla_\perp^2 = \partial/\partial x^2$ is the transverse Laplacian operator, and F_0 is a function to be evaluated by the boundary condition given by (6.78).

The optical path length P along a ray joining two points at z = 0 and z > 0 can be written as

$$P = \int_0^z L dz = \frac{1}{l_0}\int_0^z n^2(x) dz = \frac{n_s^2 z}{l_0} + \frac{(n_0^2 - n_s^2)}{l_0}\int_0^z \frac{dz}{1 + \sinh^2(\alpha x)} \tag{6.81}$$

where Eqs. (1.52a), (6.1), and (6.3) have been used. Substituting Eq. (6.11) into (6.81) and performing the integration with respect to τ, we have

$$P = \frac{n_s^2 z}{l_0} + \frac{\sqrt{n_0^2 - n_s^2}}{\alpha}\left[\tan^{-1}\left(\sqrt{1 + A^2}\tan\tau\right) - \tan^{-1}\left(\sqrt{1 + A^2}\tan\tau_0\right)\right] \tag{6.82}$$

To obtain an explicit dependence of P on the hyperbolic variable u, we can rewrite Eq. (6.82) as

$$P = \frac{n_s^2 z}{l_0} + \frac{\sqrt{n_0^2 - n_s^2}}{\alpha}\left\{\tan^{-1}\left[u\left(1 + \frac{\alpha^2\gamma^2(1 + u^2)}{\dot{u}^2}\right)^{1/2}\right] - \tan^{-1}\left[u_0\left(1 + \frac{\alpha^2\gamma^2(1 + u_0^2)}{\dot{u}_0^2}\right)^{1/2}\right]\right\} \tag{6.83}$$

Equation (6.83) has been derived taking into account the following relationships

$$1 + A^2 = 1 + u^2 + \frac{\dot{u}^2}{\alpha^2\gamma^2} = 1 + \dot{u}_0^2 + \frac{\dot{u}_0^2}{\alpha^2\gamma^2} \tag{6.84}$$

$$\tan\tau_{(0)} = \frac{u_{(0)}}{\dot{u}_{(0)}}\alpha\gamma \tag{6.85}$$

Likewise, Eq. (6.18a) can be expressed as

$$\dot{u}_0 = \frac{u(z) - u_0 H_f(z)}{H_a(z)} \tag{6.86}$$

and inserting this equation into (6.18b), we obtain

$$\dot{u}(z) = \frac{u(z)\dot{H}_a(z) - u_0}{H_a(z)} \tag{6.87}$$

where Eq. (6.20) has been used.

Then, from Eqs. (6.86–6.87) it follows that the optical path length is also written as

$$\begin{aligned} P = \frac{n_s^2 z}{l_0} + \frac{\sqrt{n_0^2 - n_s^2}}{\alpha}\Bigg\{ &\tan^{-1}\left[\frac{u}{u\dot{H}_a(z) - u_0}\sqrt{\left[u\dot{H}_a(z) - u_0\right]^2 + \alpha^2\gamma^2 H_a^2(z)(1+u^2)}\right] \\ &- \tan^{-1}\left[\frac{u_0}{u - u_0 H_f(z)}\sqrt{[u - u_0 H_f(z)]^2 + \alpha^2\gamma^2 H_a^2(z)(1+u_0^2)}\right]\Bigg\} \end{aligned} \tag{6.88}$$

P is expressed, in a more compact form, as

$$P = \frac{n_s^2 z}{l_0} + \frac{\sqrt{n_0^2 - n_s^2}}{\alpha}\tan^{-1}\left[\frac{\sqrt{u^2 + u_0^2 - 2uu_0 H_f(z) + \alpha^2\gamma^2 H_a^2(z)}}{uu_0 + H_f(z)}\right] \tag{6.89}$$

where trigonometric relationships and

$$\begin{aligned} \left[u\dot{H}_a(z) - u_0\right]^2 + \alpha^2\gamma^2 H_a^2(z)(1+u^2) &= [u - u_0 H_f(z)]^2 + \alpha^2\gamma^2 H_a^2(z)(1+u_0^2) \\ &= u_0^2 + u^2 - 2uu_0 H_f(z) + \alpha^2\gamma^2 H_a^2(z) \end{aligned} \tag{6.90}$$

have been used.

On the other hand, taking into account that

$$\nabla_\perp^2 P = \frac{\partial^2 P}{\partial x^2} = l_0\frac{\partial \dot{u}}{\partial u} = l_0\frac{\dot{H}_a(z)}{H_a(z)} \tag{6.91}$$

the amplitude of the kernel function is given by

$$\exp\left\{-\frac{1}{2l_0}\int_0^z \nabla_\perp^2 S dz'\right\} = \frac{1}{H_a^{1/2}(z)} \tag{6.92}$$

Evaluation of F_0 is carried out from condition (6.78), and by using the stationary phase method gives

$$F_0 = \left[\frac{V\alpha}{i2\pi\gamma\sqrt{1+u_0^2}}\right]^{1/2} \tag{6.93}$$

where Eq. (6.59) has been used.

Therefore, insertion of Eqs. (6.89) and (6.92–6.93) into Eq. (6.80) provides the following expression for the kernel function in a planar waveguide with HS profile [6.21]

$$\begin{aligned} K(u,u_0;z) = {} & \left[\frac{V\alpha}{i2\pi\gamma H_a(z)\sqrt{1+u_0^2}}\right]^{1/2} \exp\left\{i\frac{kn_s^2}{l_0}z\right\} \\ & \cdot \exp\left\{iV\left[\tan^{-1}\left(\frac{\sqrt{u^2+u_0^2-2uu_0H_f(z)+\alpha^2\gamma^2H_a^2(z)}}{uu_0+H_f(z)}\right)\right]\right\} \end{aligned} \tag{6.94}$$

K is now defined in the coordinate system u–z, and under transformation (6.9) the complex amplitude distribution becomes

$$\psi(u,z) = \frac{1}{\alpha}\int_{-\infty}^{\infty}\frac{K(u,u_0;z)\psi_0(u_0)}{\sqrt{1+u_0^2}}du_0 \tag{6.95}$$

Note that the kernel function given by Eq. (6.94) can be parameterized by the axial and field rays connecting the position of a ray at two transverse planes in the coordinate system u–z. Likewise, the kernel function represents the complex amplitude distribution produced by a source located at the input of the medium. Thus a source emits waves whose form, within this medium, is given by the following expression (if we choose the location $x_0 = u_0 = 0$)

$$K(u,0;z) = \left[\frac{V\alpha}{i2\pi\gamma H_a(z)}\right]^{1/2} \exp\left\{i\frac{kn_s^2}{l_0}z\right\}\exp\left\{iV\left[\tan^{-1}\left(\frac{\sqrt{u^2+\alpha^2\gamma^2H_a^2(z)}}{H_f(z)}\right)\right]\right\} \tag{6.96}$$

Finally, in most cases a paraxial approximation of the kernel function can be used. For a region very close to the z–axis, u and $\dot{u}$ are small quantities. Therefore the paraxial expression of the kernel function can be easily derived if the optical path length is written as

$$\begin{aligned} P = \frac{n_s^2 z}{l_0} + \frac{\sqrt{n_0^2-n_s^2}}{\alpha}\Bigg\{ & \tan^{-1}\left[\frac{u\alpha\gamma}{\dot{u}}\left[1+u^2+\left(\frac{\dot{u}}{\alpha\gamma}\right)^2\right]^{1/2}\right] \\ & - \tan^{-1}\left[\frac{u_0\alpha\gamma}{\dot{u}_0}\left[1+u_0^2+\left(\frac{\dot{u}_0}{\alpha\gamma}\right)^2\right]^{1/2}\right]\Bigg\} \end{aligned} \tag{6.97}$$

where Eqs. (6.82) and (6.84–6.85) have been used.

Expanding Eq. (6.97) in a series yields

$$
\begin{aligned}
P &\cong \frac{n_s^2 z}{l_0} + \frac{\sqrt{n_0^2 - n_s^2}}{\alpha}\left\{\tau - \tau_0 + \frac{1}{2\alpha\gamma}(u\dot{u} - u_0\dot{u}_0) + \cdots\right\} \\
&= \frac{n_s^2 z}{l_0} + \frac{\sqrt{n_0^2 - n_s^2}}{\alpha}\left(\alpha\gamma z + \frac{\dot{H}_a(z)u^2 + H_f(z)u_0^2 - 2uu_0}{H_a(z)} + \cdots\right)
\end{aligned}
\tag{6.98}
$$

where all higher–order terms are neglected, and Eqs. (6.12) and (6.86–6.87) have been used.

Likewise for paraxial region F_0 is approximated as

$$
F_0 \cong \left[\frac{V\alpha}{i2\pi\gamma}\right]^{1/2} \tag{6.99}
$$

Thus, the paraxial kernel function is written as

$$
\begin{aligned}
K^P(u, u_0; z) &\cong \left[\frac{V\alpha}{i2\pi\gamma H_a(z)}\right]^{1/2} \exp\left\{i\frac{kn_s^2}{l_0}z\right\}\exp\{iV\alpha\gamma z\} \\
&\cdot \exp\left\{i\frac{V}{2\alpha\gamma H_a(z)}[\dot{H}_a(z)u^2 + H_f(z)u_0^2 - 2uu_0]\right\}
\end{aligned}
\tag{6.100}
$$

and the complex amplitude distribution reduces to

$$
\psi^P(u, z) = \frac{1}{\alpha}\int_{-\infty}^{\infty} K^P(u, u_0; z)\psi_0^P(u_0)du_0 \tag{6.101}
$$

6.7 Diffraction–Free and Diffraction–Limited Propagation of Light

We limit the analysis of diffraction–free and diffraction–limited propagation of light through a planar waveguide with HS profile to the simplest guided mode for which the transverse amplitude varies in hyperbolic secant. Then, we will apply the integral equation (6.95) to a field distribution at the input given by the fundamental mode, Eq. (6.73), in the waveguide, which can be expressed in terms of the hyperbolic variable as

$$
\psi_0(x_0) = \mathrm{sech}^S(\alpha x_0) = [1 + u_0^2]^{-S/2} \tag{6.102}
$$

Thus, Eq. (6.95) can be written, taking into account Eqs. (6.94) and (6.102), as

$$
\psi(u, z) = \left[\frac{V}{i2\pi\alpha\gamma H_a(z)}\right]^{1/2} \int_{-\infty}^{\infty} \theta_0(u_0)\exp\{ik\varphi(u, u_0; z)\}du_0 \tag{6.103}
$$

where

$$\theta_0(u_0) = (1+u_0^2)^{-(3/4+S/2)} \tag{6.104}$$

$$\varphi = \frac{\sqrt{n_0^2 - n_s^2}}{\alpha} \tan^{-1}\left[\frac{\sqrt{u^2 + u_0^2 - 2uu_0 H_f(z) + \alpha^2\gamma^2 H_a^2(z)}}{uu_0 + H_f(z)}\right] \tag{6.105}$$

and Eq. (6.59) has been used.

In practice, the exact solution of the integral in Eq. (6.103) is not possible. Fortunately, $k\varphi >> 1$ for the optical domain, so the exponential term of the integrand oscillates rapidly except in the neighborhood of the stationary point. For this reason, the stationary–phase method will provide a good approximation to the value of the integral [6.22–6.23]. Applying this method to Eq. (6.103), the dominant term in the asymptotic is of order $k^{-1/2}$, and the integral can be approximated by

$$\int_{-\infty}^{+\infty} \theta_0(u_0)\exp\{ik\varphi(u,u_0;z)\}du_0 \approx \left(\frac{i2\pi}{k\tilde{\varphi}_{u_0,u_0}}\right)^{1/2} \tilde{\theta}_0 \exp\{ik\tilde{\varphi}\} \tag{6.106}$$

where $\tilde{\theta}_0$ and $\tilde{\varphi}$ are the values of θ_0 and φ at the stationary point $\tilde{u}_0$, and $\tilde{\varphi}_{u_0,u_0}$ is the second derivative of φ also evaluated at $\tilde{u}_0$.

According to this method, the stationary point is given by

$$\left(\frac{\partial \varphi}{\partial u_0}\right)_{u_0=\tilde{u}_0} = 0 \tag{6.107}$$

After a long but straightforward calculation, we have

$$\tilde{u}_0 = \frac{u}{H_f(z)} \tag{6.108}$$

From Eq. (6.108) it follows that the stationary value of θ_0, φ, and the second derivative of φ at the stationary point can be written as

$$\tilde{\theta}_0 = \frac{H_f^{S+3/2}(z)}{[u^2 + H_f^2(z)]^{3/4+S/2}} \tag{6.109}$$

$$\tilde{\varphi} = \frac{\sqrt{n_0^2 - n_s^2}}{\alpha} \tan^{-1}\left[\frac{\alpha\gamma H_a(z)}{\sqrt{u^2 + H_f^2(z)}}\right] \tag{6.110}$$

$$\tilde{\varphi}_{u_0,u_0} = \frac{H_f^4(z)\sqrt{n_0^2 - n_s^2}}{\alpha^2\gamma H_a(z)[u^2 + H_f^2(z)]^{3/2}} \tag{6.111}$$

Substituting Eqs. (6.109–6.111) into Eq. (6.106), the field distribution at z, given by Eq. (6.103), becomes

$$\psi(u,z) \cong \frac{H_f^{S-1/2}(z)}{[u^2 + H_f^2(z)]^{S/2}} \exp\left\{ iV\tan^{-1}\left[\frac{\alpha\gamma H_a(z)}{\sqrt{u^2 + H_f^2(z)}} \right] \right\} \tag{6.112}$$

where Eq. (6.59) has been used.

Thus, the irradiance distribution due to the propagation of the fundamental mode through the waveguide is given by

$$I(u,z) = |\psi(u,z)|^2 \cong \frac{|H_f(z)|^{2S-1}}{[u^2 + H_f^2(z)]^S} \tag{6.113}$$

For points on the axis (u = x = 0), irradiance reduces to

$$I_0(z) \cong \frac{1}{|H_f(z)|} \tag{6.114}$$

From Eq. (6.114) it follows that at $z_m = m\pi/\alpha\gamma$ such that $|H_f(z_m)| = 1$ or $H_a(z_m) = 0$, the on–axis irradiance achieves its minimum value $I_0(z_m) = 1$. It also follows that at $z_p = (2p+1)\pi/\alpha\gamma$ such that $H_f(z_p) = 0$, the on–axis irradiance goes to infinity.

On the other hand, the off–axis irradiance (u and x ≠ 0) at z_m becomes

$$I(u,z_m) \cong \frac{1}{[1+u^2]^S} = \operatorname{sech}^{2S}(\alpha x) \tag{6.115}$$

that is, at z_m the irradiance is a replica of the input one. Then $H_a(z_m) = 0$ is the imaging condition, as may be expected from the ray analysis. Likewise, at z_p the off–axis irradiance tends to zero. By taking into account its on–axis behavior, it follows that the irradiance at z_p becomes a one–dimensional Dirac delta function with $H_f(z_p) = 0$ or the focusing condition, as expected. In short, imaging and focusing conditions for the propagation of the fundamental mode in this planar waveguide are given by the zeros of functions H_a and H_f, respectively [6.24].

Figure 6.16 shows the non–paraxial diffraction–free propagation of the fundamental mode through a planar waveguide with hyperbolic secant refractive index profile for S = 0.5 and at several values of z in the interval. The irradiance distribution versus u and x gradually narrows as z increases.

Figure 6.17 represents the irradiance distribution for S = 0.5 in the neighborhood of $z = \pi/2\alpha\gamma$ where it becomes a one–dimensional delta function corresponding to the first zero of the field ray. Level curves of irradiance on the x–z plane in the interval $[0, \pi/\alpha\gamma]$ show this behavior at $z = \pi/2\alpha\gamma$. At $z = \pi/\alpha\gamma$, where the first zero of the axial ray is obtained, they show the recovery of the input irradiance.

Figure 6.18 depicts the irradiance distribution at $z = \pi/3\alpha\gamma$ versus hyperbolic and real coordinates u and x for several values of z. The irradiance is normalized to its maximum value achieved at the center in all cases, and the width of the irradiance distribution narrows as S increases.

Until now we have neglected the finite size of the planar waveguide. We shall now analyze the effect of this limitation on the fundamental mode propagation in this kind of planar waveguide with an aperture of size 2a.

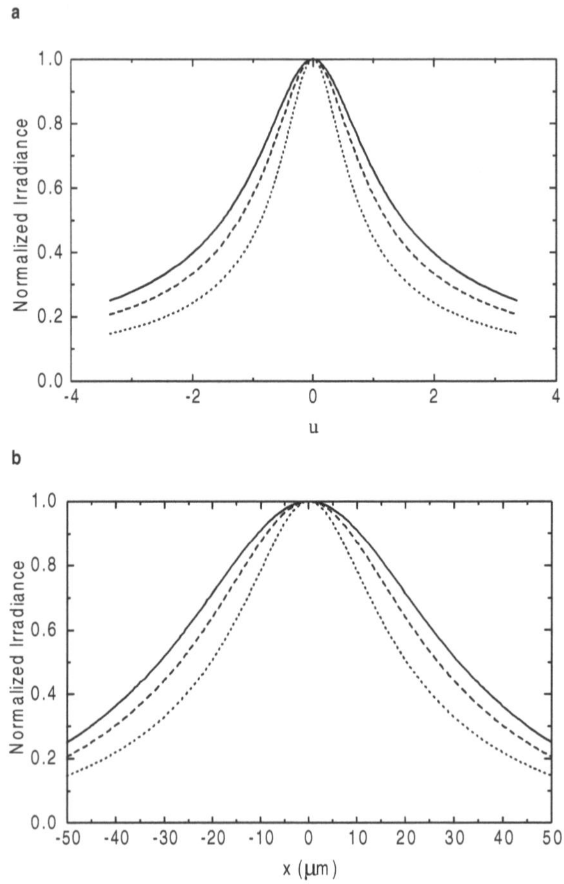

Fig. 6.16. Non–paraxial diffraction–free propagation of the fundamental mode. Irradiance distribution for S = 0.5 **a** versus hyperbolic coordinate u; **b** versus real coordinate x at waveguide length $z = \pi / 6\alpha\gamma$ (solid line), $\pi / 4\alpha\gamma$ (dashed line), $\pi / 3\alpha\gamma$ (dotted line). Calculations have been made for $n_0 = 1.5$, $n_s = 1.4476$, $\lambda = 1.31\,\mu m$, $\gamma = 0.058$, and $\alpha = 38.5\,mm^{-1}$.

a

b

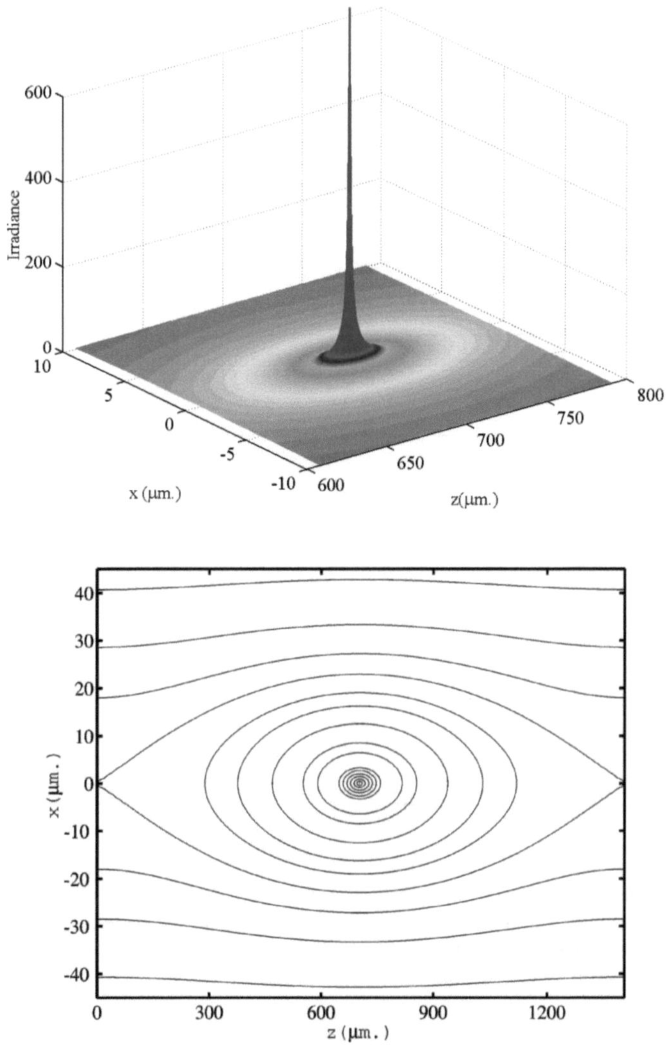

Fig. 6.17. Non–paraxial diffraction–free propagation of the fundamental mode. Irradiance distribution for S = 0.5 **a** in the neighborhood of $z = \pi/2\alpha\gamma$; **b** level curves of the irradiance on the x–z plane in the interval [0, $\pi/\alpha\gamma$]. Calculations have been made for $n_0 = 1.5$, $n_s = 1.4476$, $\lambda = 1.31\,\mu m$, $\gamma = 0.058$, and $\alpha = 38.5\,mm^{-1}$.

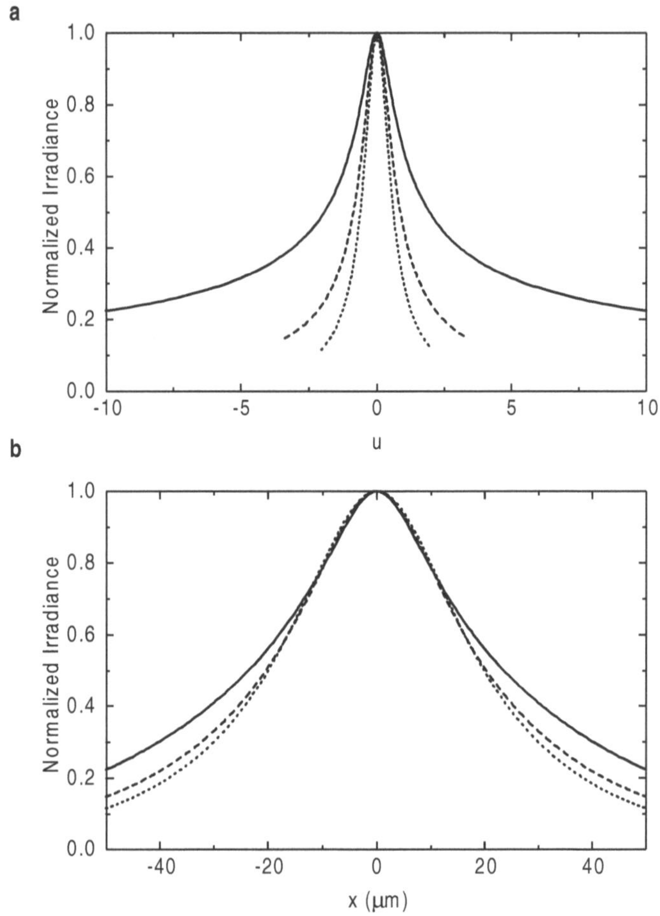

Fig. 6.18. Irradiance distribution at $z = \pi / 3\alpha\gamma$ for S = 0.25 (solid line), 0.5 (dashed line), and 0.75 (dotted line) versus **a** hyperbolic coordinate u; and **b** real coordinate x. Calculations have been made for $n_0 = 1.5$, $n_s = 1.4476$, $\lambda = 1.31\,\mu m$, $\gamma = 0.058$, and $\alpha = 60\,mm^{-1}$ (S = 0.25), $\alpha = 38.5\,mm^{-1}$ (S = 0.5), and $\alpha = 29.3\,mm^{-1}$ (S = 0.75).

If we take the aperture of the waveguide into account, Eq. (6.103) becomes

$$\psi^{dl}(u,z) = \left[\frac{V}{i2\pi\alpha\gamma H_a(z)}\right]^{1/2} \int_{-u_0(a)}^{u_0(a)} \theta_0(u_0) \exp\{ik\varphi(u,u_0;z)\} du_0 \qquad (6.116)$$

The stationary phase method, once again, provides a good approximation to the value of the integral in Eq. (6.116), but the dominant terms in the asymptotic expansion depend on the location of the stationary point with respect to the interval of integration [6.22–6.23].

For $|\tilde{u}_0| < |u_0(a)|$, $|u| < |u_0(a)H_f(z)|$ (geometrically illuminated region at z), and the stationary point is within the interval of integration. Applying the stationary–phase method to the integral in Eq. (6.116) yields a dominant term in the asymptotic expansion of the integral of order $k^{-1/2}$, and the next term is found from the integration near the endpoints of the interval of integration. Then the integral in Eq. (6.116) can be approximated by

$$\int_{-u_0(a)}^{u_0(a)} \theta_0(u_0)\exp\{ik\varphi(u,u_0;z)\}du_0 \cong \left(\frac{i2\pi}{k\tilde{\varphi}_{u_0,u_0}}\right)\tilde{\theta}_0 \exp\{ik\tilde{\varphi}\} + \frac{1}{ik}\left[\frac{\theta_0[u_0(a)]\exp\{ik\varphi[u,u_0(a);z]\}}{\varphi_{u_0(a)}} - \frac{\theta_0[-u_0(a)]\exp\{ik\varphi[u,-u_0(a);z]\}}{\varphi_{-u_0(a)}}\right] \tag{6.117}$$

where $\tilde{\theta}_0$, $\tilde{\varphi}$, and $\tilde{\varphi}_{u_0,u_0}$ are given by Eqs. (6.109–6.111) and $\theta_0[\pm u_0(a)]$, $\varphi[u,\pm u_0(a);z]$, and $\varphi_{\pm u_0(a)}$ are the values of θ_0, φ, and the first derivative of φ at the endpoints of integration.

Substituting Eqs. (6.109–6.111) into Eq. (6.117) transforms the field distribution at z given by Eq. (6.116) into

$$\begin{aligned}\psi^{dl}(u,z) \cong \psi(u,z) + &\left[\frac{i}{2\pi k\gamma H_a(z)\sqrt{n_0^2-n_s^2}}\right]^{1/2} \frac{1}{\left[u^2-u_0^2(a)H_f^2(z)\right]\left[1+u_0^2(a)\right]^{S/2-1/4}} \\ &\cdot\Big\{\left[u+u_0(a)H_f(z)\right]\left[u^2+u_0^2(a)-2uu_0(a)H_f(z)+\alpha^2\gamma^2H_a^2(z)\right]^{1/2} \\ &\cdot\exp\{ik\varphi[u,u_0(a);z]\} - \left[u-u_0(a)H_f(z)\right] \\ &\cdot\left[u^2+u_0^2(a)+2uu_0(a)H_f(z)+\alpha^2\gamma^2H_a^2(z)\right]^{1/2}\exp\{ik\varphi[u,-u_0(a);z]\}\Big\}\end{aligned} \tag{6.118}$$

where the first term in Eq. (6.118) is the field distribution for diffraction–free propagation.

After a long but straightforward calculation, the irradiance distribution in the geometrically illuminated region is given by

$$\begin{aligned}I^{dl}(u,z) = |\psi^{dl}(u,z)| \cong\; & I(u,z) + B_+^2 + B_-^2 - 2I^{1/2}(u,z)B_+\cos\left(kC_- - \pi/4\right) \\ & + 2I^{1/2}(u,z)B_-\cos\left(kC_+ - \pi/4\right) - 2B_+B_-\cos(kD)\end{aligned} \tag{6.119}$$

where the first term in Eq. (6.119) is the irradiance distribution for diffraction–free propagation,

$$B_\pm^2 = \frac{u^2+u_0^2(a)\pm 2uu_0(a)H_f(z)+\alpha^2\gamma^2H_a^2(z)}{2\pi k\gamma|H_a(z)|\sqrt{n_0^2-n_s^2}\,[u\pm u_0(a)H_f^2(z)]^2[1+u_0^2(a)]^{S-1/2}} \tag{6.120}$$

$$C_{\pm} = \tilde{\varphi} - \varphi[u, \pm u_0(a); z] = \frac{\sqrt{n_0^2 - n_s^2}}{\alpha} \cdot$$

$$\tan^{-1}\left[\frac{[H_f(z) \pm uu_0(a)]\alpha\gamma H_a(z) - \{[u^2 + H_f^2(z)][u^2 + u_0^2(a) \pm 2uu_0(a)H_f(z) + \alpha^2\gamma^2 H_a^2(z)]\}^{1/2}}{[H_f(z) \pm uu_0(a)][u^2 + H_f^2(z)]^{1/2} + \alpha\gamma H_a(z)[u^2 + u_0^2(a) \pm 2uu_0(a)H_f(z) + \alpha^2\gamma^2 H_a^2(z)]^{1/2}}\right] \tag{6.121}$$

with plus for B_+ or C_+, and minus for B_- and C_- in Eqs. (6.119–6.121); and

$$D = \varphi[u, u_0(a); z] - \varphi[u, -u_0(a); z] = \frac{\sqrt{n_0^2 - n_s^2}}{\alpha}$$

$$\tan^{-1}\left[\frac{[H_f(z) - uu_0(a)][u^2 + u_0^2(a) - 2uu_0(a)H_f(z) + \alpha^2\gamma^2 H_a^2(z)]^{1/2}}{H_f^2(z) - u^2u_0^2(a) + \left\{[u^2 + u_0^2(a) + \alpha^2\gamma^2 H_a^2(z)]^2 - 4u^2u_0^2(a)H_f^2(z)\right\}^{1/2}}\right.$$

$$\left. - \frac{[H_f(z) + uu_0(a)][u^2 + u_0^2(a) + 2uu_0(a)H_f(z) + \alpha^2\gamma^2 H_a^2(z)]^{1/2}}{H_f^2(z) - u^2u_0^2(a) + \left\{[u^2 + u_0^2(a) + \alpha^2\gamma^2 H_a^2(z)]^2 - 4u^2u_0^2(a)H_f^2(z)\right\}^{1/2}}\right] \tag{6.122}$$

Consequently, for a geometrically illuminated region, the irradiance at z may be regarded as the sum of the irradiance when the finite extent of the planar waveguide has been neglected and another terms due to the diffractional effect of the aperture.

For $|\tilde{u}_0| = |u_0(a)|$, it follows that $|u| = |u_0(a)H_f(z)|$ (the edge of geometrical shadow), and the stationary point is at the upper or lower endpoint of the interval of integration. Then, the integral tends to zero more slowly than k^{-1} and can be approximated by

$$\int_{-u_0(a)}^{u_0(a)} \theta_0(u_0)\exp\{ik\varphi(u, u_0; z)\}du_0 \cong \left(\frac{i\pi}{2k\varphi_{u_0(a),u_0(a)}}\right)^{1/2} \theta_0[u_0(a)]\exp\{ik\varphi[u, u_0(a); z]\} \tag{6.123}$$

where $\varphi_{u0(a),u0(a)}$ is the second derivative of φ evaluated at $u_0(a)$.

In this case, the complex amplitude distribution is expressed as

$$\psi^{dl}(u, z) \cong \frac{H_f^{S-1/2}(z)}{2[u^2 + H_f^2(z)]^{S/2}} \exp\left\{iV\tan^{-1}\left[\frac{\alpha\gamma H_a(z)}{\sqrt{u^2 + H_f^2(z)}}\right]\right\} \tag{6.124}$$

and the irradiance is given by

$$I^{dl}(u, z) \cong \frac{|H_f(z)|^{2S-1}}{4[u^2 + H_f^2(z)]^{S/2}} = \frac{I(u, z)}{4} \tag{6.125}$$

From Eqs. (6.125) and (6.113) it follows that the irradiance at the edge of geometrical shadow is one–quarter of the corresponding value for diffraction–free propagation.

For $|\tilde{u}_0| > |u_0(a)|$, it follows that $|u| > |u_0(a)H_f(z)|$ (geometrical shadow), and the stationary–phase method is outside the interval of integration. Then the principal term in the asymptotic expansion of the integral is of order k^{-1} and can be approximated by

$$\int_{-u_0(a)}^{u_0(a)} \theta_0(u_0)\exp\{ik\varphi(u,u_0;z)\}du_0 \cong \frac{1}{ik}\left[\frac{\theta_0[u_0(a)]\exp\{ik\varphi[u,u_0(a);z]\}}{\varphi_{u_0(a)}} - \frac{\theta_0[-u_0(a)]\exp\{ik\varphi[u,-u_0(a);z]\}}{\varphi_{-u_0(a)}}\right] \tag{6.126}$$

The complex amplitude distribution is given by

$$\begin{aligned}\psi^{dl}(u,z) \cong & \left[\frac{i}{2\pi k\gamma H_a(z)\sqrt{n_0^2-n_s^2}}\right]^{1/2}\frac{1}{\left[u^2-u_0^2(a)H_f^2(z)\right]\left[1+u_0^2(a)\right]^{S/2-1/4}} \\ & \cdot\Big\{\left[u+u_0(a)H_f(z)\right]\left[u^2+u_0^2(a)-2uu_0(a)H_f(z)+\alpha^2\gamma^2H_a^2(z)\right]^{1/2} \\ & \cdot\exp\{ik\varphi[u,u_0(a);z]\}-\left[u-u_0(a)H_f(z)\right] \\ & \cdot\left[u^2+u_0^2(a)+2uu_0(a)H_f(z)+\alpha^2\gamma^2H_a^2(z)\right]^{1/2}\exp\{ik\varphi[u,-u_0(a);z]\}\Big\}\end{aligned} \tag{6.127}$$

For the irradiance distribution we obtain

$$\begin{aligned}I^{dl}(u,z) \cong & \frac{1}{\pi\alpha V}\left\{\frac{\alpha^2\gamma|H_a(z)|\left[u^2+u_0^2(a)H_f^2(z)\right]}{\left[u^2-u_0^2(a)H_f^2(z)\right]\left[1+u_0^2(a)\right]^{S-3/2}}+\frac{1}{\gamma|H_a(z)|\left[1+u_0^2(a)\right]^{S-1/2}}\right. \\ & \left.\cdot\left[1-\frac{\left[\left[u^2+u_0^2(a)+\alpha^2\gamma^2H_a^2(z)\right]^2-4u^2u_0^2H_f^2(z)\right]^{1/2}}{u^2-u_0^2(a)H_f^2(z)}\cos(kD)\right]\right\}\end{aligned} \tag{6.128}$$

where D is given by Eq. (6.122).

Thus for the geometrical shadow, the irradiance at z can be regarded as being due to diffractional effects. At $z_p = (2p+1)\pi/2\alpha\gamma$ such that $H_f(z_p) = 0$, Eq. (6.128) reduces to

$$I^{dl}(u,z_p) \cong \frac{2[1+u^2+u_0^2(a)]}{\pi u^2 V[1+u_0^2(a)]^{S-1/2}}\sin^2\left[V\tan^{-1}\left[\frac{uu_0(a)}{\sqrt{1+u^2+u_0^2(a)}}\right]\right] \tag{6.129}$$

Figure 6.19 shows the diffraction–limited propagation of the fundamental mode for $S = 0.5$. The level curves of irradiance on the x–z plane in the length interval $[0, 4\pi/\alpha\gamma]$ are presented in Fig. 6.19a. The irradiance has principal maxima in the

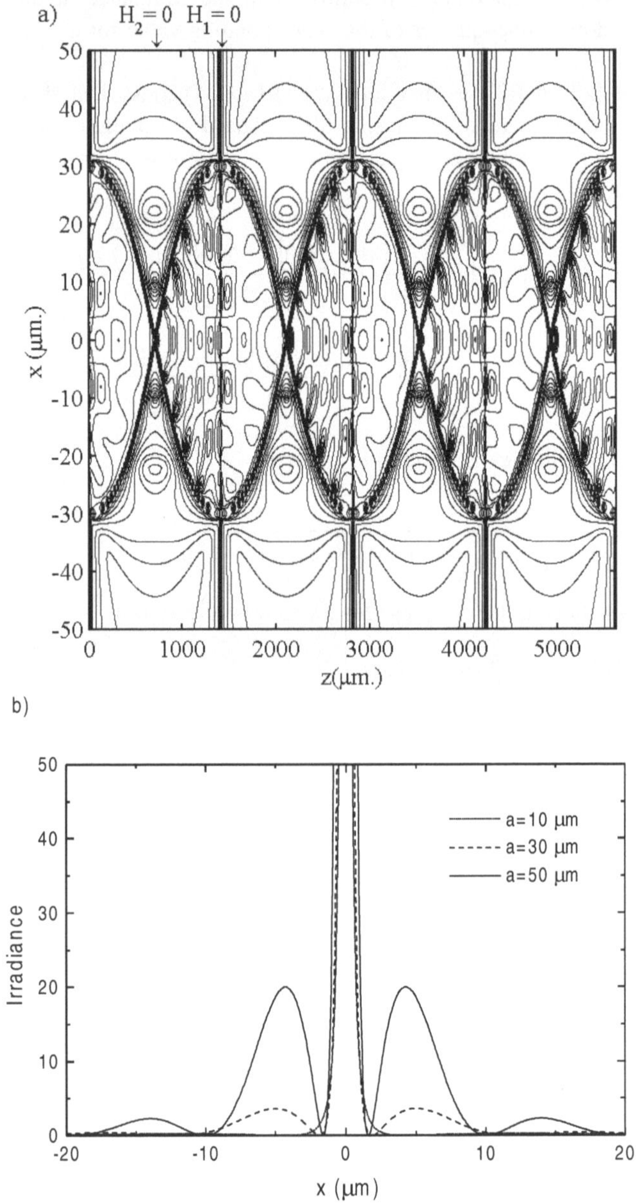

Fig. 6.19. Non–paraxial diffraction–limited propagation of the fundamental mode. **a** Level curves of irradiance on the x–z plane along the interval [0, $4\pi/\alpha\gamma$] for $a = 30\,\mu m$. **b** Enlarged central region of diffraction pattern at $z = \pi/2\alpha\gamma$ for $a = 10, 30, 50\,\mu m$. In both cases S = 0.5 ($\alpha = 38.5\ mm^{-1}$).

illuminated region near the values of u given by the stationary point. Separation between the maxima decreases with increasing length of the planar waveguide as a result of the dependence of the stationary point on H_f. At $z = \pi / 2\alpha\gamma$, where the first zero of H_f is obtained, the maxima are superimposed. There is a reverse behavior in the interval $(\pi / 2\alpha\gamma, \pi / \alpha\gamma)$, and at $z = \pi / \alpha\gamma$ the first zero of H_a is reached. Likewise, it is also of interest to examine the dependence of diffraction–limited propagation on the aperture. Figure 6.19b depicts an enlarged central region of the diffraction pattern at $z = \pi / 2\alpha\gamma$ to show the variation of the width of the principal maximum and of the position of the minima with semiapertures [6.25].

Figure 6.20 represents the variation of the positions of the maxima with S at a given length. The separation between maxima decreases as S increases. In Figs.6.19 and 6.20 calculations were made for $n_0 = 1.5$, $n_s = 1.4476$, $\lambda = 1.31\,\mu m$, and $\gamma = 0.058$.

The above results, obtained by the stationary–phase method, can be generalized to any field distributions at the input that correspond to guided modes in the planar waveguide with HS profile. Finally, diffraction–free and diffraction–limited propagation of light in the paraxial region can also be evaluated and analyzed in the same way by using Eqs. (6.100–6.101) [6.26].

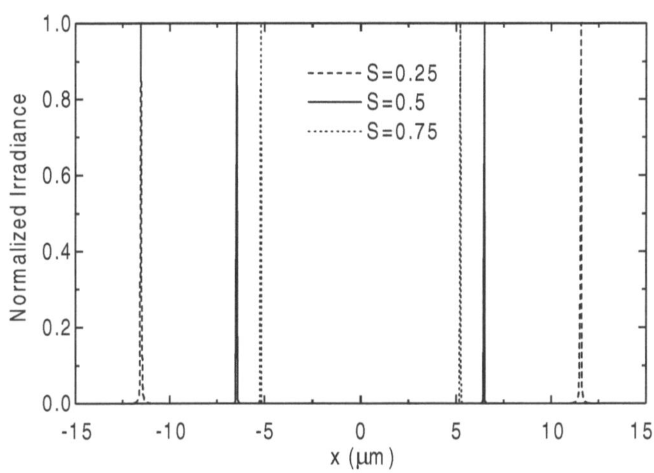

Fig. 6.20. Irradiance distribution at $z = \pi / 2.1\alpha\gamma$ versus x for $a = 50\,\mu m$. Calculations have been made for S = 0.25 ($\alpha = 60\,mm^{-1}$), S = 0.5 ($\alpha = 38.5\,mm^{-1}$) and S = 0.75 ($\alpha = 29.3\,mm^{-1}$).

7 The Talbot Effect in GRIN Media

7.1 Introduction

The Talbot effect, consisting of a series of self–images in planes beyond a grating illuminated by coherent light, is well known in optics and has received wide attention. Fundamental properties and many practical applications of the Talbot effect in a homogeneous medium have been considered [7.1–7.10]. The Talbot effect has also been studied in the context of atom optics because of similarities between the Schrödinger and the paraxial wave equations [7.11–7.12]. Likewise, the self–imaging phenomenon, in general, can be treated as a superposition of a proper set of modes in either free space or inhomogeneous media [7.13].

This chapter generalizes the self–imaging phenomenon to a GRIN medium with a transverse parabolic index profile modulated by an axial index when a periodic object located at the input of the medium is illuminated by coherent light. The analysis will be restricted to the one–dimensional transverse case, but extension to the two–dimensional case is straightforward.

Theoretical aspects of light propagation will be investigated assuming strictly periodic objects of infinite dimensions illuminated by coherent light. Such a theoretical description is sufficient for many studies. In particular, integer and fractional Talbot effect as well as an analogy with an apparent lens of multifocal distances will be derived and discussed.

However, it is important to realize the effects arising as a result of the departure from the theoretical model. In this way, the influence of the displacement of the source and of the finite transverse dimension of the hybrid optical structures formed by periodic objects and GRIN media will be studied in order to get a more realistic description of light propagation through these hybrid systems. The results presented in this chapter, can have important and wide applicability in imaging for increasing or decreasing spatial frequency, in integrated optics for transverse couplers and multimode interference filters, and so on [7.14–7.17].

7.2
Light Propagation and Imaging Condition

Let us consider a GRIN medium characterized by a refractive index profile given by Eq. (1.69a). To evaluate the complex amplitude distribution in this GRIN medium, we assume that a one–dimensional periodic object of infinite dimension located at the input is illuminated by a coherent nonuniform beam (Fig. 7.1). The one–dimensional periodic object is represented as

$$T(x_0) = \sum_{q=-\infty}^{\infty} a_q \exp\left\{-i\frac{2\pi q x_0}{p}\right\} \tag{7.1}$$

where p is the spatial period, and a_q is the amplitude of the qth harmonic.

When the hybrid structure formed by the periodic object and the GRIN medium is illuminated by a coherent non–uniform beam, the complex amplitude distribution on the periodic object located at z = 0 can be written as

$$\psi(x_0) = T(x_0)\psi_0(x_0) \tag{7.2}$$

where

$$\psi_0(x_0) = \left[\frac{w_0}{w(0)}\right]^{1/2} \exp\{i\varphi(0)\}\psi[x_0; U(0)] \tag{7.3}$$

is the complex amplitude distribution due to a Gaussian illumination of wavelength λ,

$$\psi[x_0; U(0)] = \exp\left\{i\frac{\pi U(0) x_0^2}{\lambda}\right\} \tag{7.4}$$

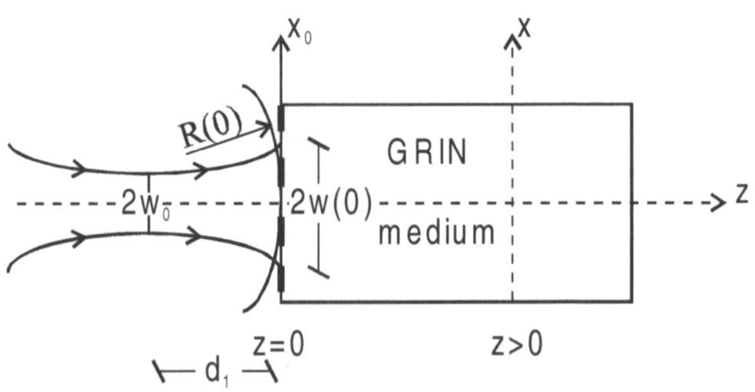

Fig. 7.1. Geometry for the evaluation of the complex amplitude distribution in a GRIN medium due to a periodic object located at z=0 and illuminated by a Gaussian beam.

and U(0) and φ(0) are given by Eqs. (4.2–4.3), respectively.

Substituting Eqs. (7.2–7.4) into this equation

$$\psi(x;z) = \int_{-\infty}^{+\infty} \psi(x_0)K(x,x_0;z)dx_0 \tag{7.5}$$

where K is the one–dimensional kernel function of the GRIN medium. By integrating, we arrive at the following complex amplitude distribution at z > 0

$$\psi(x;z) = \left[\frac{w_0}{w(0)G(z)}\right]^{1/2} \exp\{i\varphi(z)\}\exp\left\{i\frac{\pi U(z)}{\lambda}x^2\right\}\sum_{q=-\infty}^{\infty} a_q \exp\left\{-i\frac{2\pi q}{pG(z)}x\right\} \exp\left\{-i\frac{\pi\lambda q^2 H_a(z)}{n_0 p^2 G(z)}\right\} \tag{7.6}$$

where U(z), φ(z), and G(z) are given by Eqs. (4.8–4.10), respectively.

Equation (7.6) is the central result of the present analysis and represents the light propagation as a superposition of the object diffraction orders through the GRIN medium.

The second exponential term of the summation in Eq. (7.6) is of basic importance to imaging. Now we analyze the imaging condition that this term cancels. At the condition given by Eq. (2.25), the phase term becomes unity, and Eq. (7.6) reduces to

$$\psi(x;z_m) = \left[\frac{w_0}{w(z_m)}\right]^{1/2} \exp\{i\varphi(z_m)\}\exp\left\{i\frac{\pi U(z_m)}{\lambda}x^2\right\}\sum_{q=-\infty}^{\infty} a_q \exp\left\{-i\frac{2\pi q}{p(z_m)}x\right\} \tag{7.7}$$

where the beam half–width, complex curvature, and object period at z_m are given, respectively, by

$$w(z_m) = w(0)H_f(z_m) \tag{7.8}$$

$$U(z_m) = U(0)\frac{\dot{H}_a(z_m)}{H_f(z_m)} \tag{7.9}$$

$$p(z_m) = pH_f(z_m) \tag{7.10}$$

When Eq. (7.7) is compared with Eqs. (7.1–7.2) and Eqs. (7.3–7.4) are taken into account, it follows that the complex distribution at distances z_m is a replica of the complex amplitude at z = 0, as the well–known imaging condition $H_a(z_m) = 0$ for a GRIN medium is satisfied.

7.3 The Integer Talbot Effect

In addition to the imaging phenomenon, periodic repetition along the z–axis of the lateral complex amplitude distribution occurs for the Talbot condition. The summation in Eq. (7.6) can be rewritten as

$$\sum_{q=-\infty}^{\infty} a_q \exp\left\{-\frac{\pi\lambda q^2\,\mathrm{Im}[G(z)]}{n_0 p^2 |G(z)|^2} H_a(z)\right\} \exp\left\{-\frac{2\pi q\,\mathrm{Im}[G(z)]}{p|G(z)|^2} x\right\}$$
$$\exp\left\{-i\frac{\pi\lambda q^2\,\mathrm{Re}[G(z)]}{n_0 p^2 |G(z)|^2} H_a(z)\right\} \exp\left\{-i\frac{2\pi q\,\mathrm{Re}[G(z)]}{p|G(z)|^2} x\right\} \tag{7.11}$$

where $|G(z)|$ is the modulus of G(z) given by Eq. (4.15), and Re[G(z)] and Im[G(z)] are the real and imaginary parts of G(z), that is

$$\mathrm{Re}[G(z)] = \frac{H_a(z)}{n_0 R(0)} + H_f(z) \tag{7.12a}$$

$$\mathrm{Im}[G(z)] = \frac{H_a(z)}{z_R} \tag{7.12b}$$

where the Rayleigh range z_R is given by Eq. (4.14).

The first two exponential terms and the next two phase terms of Eq. (7.11) describe amplitude and phase changes of diffraction orders along the axial and transverse directions, respectively. In particular, the third term represents the phase changes of diffraction orders with axial distances. When this term becomes unity, all diffraction orders are cophasic and reinforce at distances z_v, fulfilling the relation

$$\frac{\lambda\,\mathrm{Re}[G(z_v)]}{n_0 p^2 |G(z_v)|^2} H_a(z_v) = 2v$$

or

$$\frac{2vp^2}{\lambda} = \frac{\mathrm{Re}[G(z_v)]}{n_0 |G(z_v)|^2} H_a(z_v) \tag{7.13}$$

called the Talbot condition, where v is an integer referred to as the self–image number [7.18].

In this case, the last term of Eq. (7.11), which describes the phase changes of the diffraction orders along the lateral direction, can be expressed as

$$\exp\left[-i\frac{2\pi q}{p(z_v)} x\right] \tag{7.14}$$

where the period of self–images

$$p(z_\nu) = pM_t(z_\nu) \tag{7.15}$$

carries information about the transverse magnification of a periodic object due to Gaussian illumination, given by

$$M_t(z_\nu) = \frac{|G(z_\nu)|^2}{\mathrm{Re}[G(z_\nu)]} = \frac{w^2(z_\nu)}{w^2(0)\,\mathrm{Re}[G(z_\nu)]} \tag{7.16}$$

where Eq. (4.13) has been used.

Note that the factor 2 in Eq. (7.13) can be omitted. In such a case, when ν is an odd integer, images present a transverse shift with respect to the object as is proved in Sect. 7.5.

Equation (7.13), which determines the Talbot condition for a GRIN medium, may also be expressed as

$$\frac{2\nu p^2}{\lambda d_0^2} = \frac{1}{R(0)} - \frac{1}{z'_\nu} \tag{7.17}$$

where

$$z'_\nu = R(0) + z_\nu = \frac{|G(z_\nu)|^2}{\mathrm{Re}[G(z_\nu)]} R(0) \tag{7.18a}$$

$$z_\nu = \frac{|G(z_\nu)|^2 - \mathrm{Re}[G(z_\nu)]}{\mathrm{Re}[G(z_\nu)]} R(0) \tag{7.18b}$$

$$d_0^2 = \frac{H_a(z_\nu)}{n_0 z_\nu} R^2(0) \tag{7.18c}$$

Therefore, the integer Talbot effect in GRIN media can be considered the equivalent effect of an apparent lens of multifocal length

$$f_\nu = \frac{\lambda d_0^2}{2\nu p^2} \tag{7.19}$$

situated at the curvature center of the Gaussian beam that illuminates a periodic object, as shown in Fig. 7.2a.

The transverse magnification of the apparent lens is now given by the ratio of the distances of the observation and the object planes from the curvature center of the beam, that is

$$\frac{z_\nu'}{R(0)} = \frac{|G(z_\nu)|^2}{\mathrm{Re}[G(z_\nu)]} \tag{7.20}$$

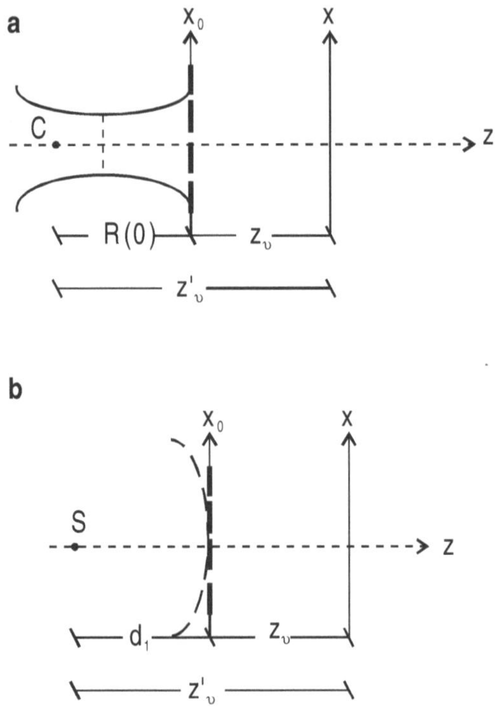

Fig. 7.2. Equivalent optical system for the Talbot effect in a GRIN medium under **a** divergent Gaussian and **b** uniform illumination.

which coincides with Eq. (7.16).

If the incident Gaussian beam has its waist located on the input, in other words, if the periodic object is illuminated by a plane Gaussian beam, then $R(0)\to\infty$, and G(z) becomes $G_p(z)$, given by Eq. (4.17). Inserting this equation into Eq. (7.13), the Talbot condition for plane Gaussian illumination reduces to

$$\frac{2\nu p^2}{\lambda} = \frac{H_f(z_v)}{n_0 \,|\, G_p(z_v)\,|^2} H_a(z_v) \tag{7.21}$$

and the transverse magnification becomes

$$M_t^{pg}(z_v) = \frac{|\, G_p(z_v)\,|^2}{H_f(z_v)} \tag{7.22}$$

Likewise, Eq. (7.11) includes uniform illumination as a special case. For this illumination $R(0)\to d_1$, $w(0)\to\infty$, and $z_R\to\infty$. Under these conditions, G(z) becomes F(z), given by Eq. (3.92). The Talbot condition and the equivalent lens equation are now given by

$$\frac{\lambda H_a(z_v)}{n_0 p^2 F(z_v)} = 2\nu$$

or

$$\frac{2\nu p^2}{\lambda} = \frac{d_1 H_a(z_v)}{n_0 d_1 H_f(z_v) + H_a(z_v)} \tag{7.23}$$

$$\frac{2\nu p^2}{\lambda d_{u0}^2} = \frac{1}{d_1} - \frac{1}{z'_v} \tag{7.24}$$

where

$$z'_v = d_1 + z_v = d_1 F(z_v) \tag{7.25a}$$

$$z_v = [F(z_v) - 1] d_1 \tag{7.25b}$$

$$d_{u0}^2 = \frac{H_a(z_v)}{n_0 z_v} d_1^2 \tag{7.25c}$$

Equation (7.24) can be regarded as the equation of an apparent lens of multifocal distance

$$f_v^u = \frac{\lambda d_{u0}^2}{2\nu p^2} \tag{7.26}$$

located at the source illuminating the object as shown in Fig. 7.2b, and whose lateral magnification is written as

$$M_t^u(z_v) = \frac{z'_v}{d_1} = F(z_v) \tag{7.27}$$

This interpretation is in full agreement with geometrical shadow principle [7.4, 7.19].

The Talbot condition and transverse magnification of integer self–images for uniform plane illumination are obtained from Eqs. (7.23–7.24) by putting $d_1 \to \infty$, that is

$$\frac{2\nu p^2}{\lambda} = \frac{H_a(z_v)}{n_0 H_f(z_v)} \tag{7.28}$$

$$M_t^{pu}(z_v) = H_f(z_v) \tag{7.29}$$

When Eq. (7.29) is compared with Eq. (7.10), it follows that the transverse magnification is the same as for non–uniform illumination when the imaging condition is fulfilled.

Finally, for free space where $g(z) \to 0$, $n_0 = 1$, $H_a(z) \to z$, and $H_f(z) \to 1$, the Talbot condition and the equivalent lens equation become, respectively

$$\frac{2\nu p^2}{\lambda} = \frac{z_v z'_v}{R(0)}\left[\frac{w(0)}{w(z_v)}\right]^2 \tag{7.30}$$

$$\frac{2\nu p^2}{\lambda^2 R^2(0)} = \frac{1}{R(0)} - \frac{1}{z'_v} \tag{7.31}$$

with $z'_v = R(0) + z_v$ for Gaussian illumination, and

$$\frac{2\nu p^2}{\lambda} = \frac{z_v d_1}{z'_v} \tag{7.32}$$

$$\frac{2\nu p^2}{\lambda d_1^2} = \frac{1}{d_1} - \frac{1}{z'_v} \tag{7.33}$$

with $z'_v = d_1 + z_v$ for uniform illumination [7.19].

7.4 Self–Image Distances

The self–image distances correspond to the lengths z_v of a GRIN medium for which the input complex amplitude distribution is periodically repeated along the optical z–axis of this medium.

The axial localization of self–images can be obtained from the Talbot condition, if the axial and field rays are written as Eqs. (4.38–4.39). Substituting Eqs. (7.12a), (4.15), and (4.38–4.39) into Eq. (7.13), we have the following second–order equation

$$a u^2(z_v) + b u(z_v) + c = 0 \tag{7.34}$$

where $u(z_v)$ is given by Eq. (4.40) and

$$a = \frac{1}{g_0}\left[2\nu p^2 n_0\left(\frac{1}{n_0^2 R^2(0)} + \frac{1}{z_R^2}\right) - \frac{\lambda}{n_0 R(0)}\right] \tag{7.35}$$

$$b = \frac{4\nu p^2}{R(0)} - \lambda \tag{7.36}$$

$$c = 2\nu p^2 n_0 g_0 \tag{7.37}$$

The solution of Eq. (7.34) is expressed as

$$u(z_v) = \frac{n_0 g_0 R^2(0) z_R^2 \lambda}{4\nu p^2 [z_R^2 + n_0^2 R^2(0)] - 2\lambda R(0) z_R^2}\left[1 - \frac{4\nu p^2}{R(0)\lambda} - \sqrt{1 - \left(\frac{4 n_0 \nu p^2}{z_R \lambda}\right)^2}\right] \tag{7.38}$$

Note that the negative sign for the square root in Eq. (7.38) has been taken for recovering the initial condition $u(z_v) = 0$ for $v = 0$, which indicates the position of the periodic object at the input.

From Eq. (7.38) it follows that the real values of $u(z_v)$ are achieved as the requirement

$$\frac{4n_0 v p^2}{z_R \lambda} \le 1 \quad \text{or} \quad v \le \frac{\pi}{4}\left[\frac{w(0)}{p}\right]^2 \tag{7.39}$$

is fulfilled, where Eq. (4.14) has been used.

Then, the integer v is limited by a square relationship between the beam half–width at $z = 0$ and the period of the object [7.5]. One or more self–images are obtained as $w(0)/p \ge \sqrt{4/\pi}$.

If the periodic object is illuminated by a plane Gaussian beam, Eq. (7.38) reduces to

$$u(z_v) = \frac{g_0 \lambda z_{pR}^2}{4vp^2 n_0}\left[1 - \sqrt{1 - \left(\frac{4n_0 v p^2}{z_{pR}\lambda}\right)^2}\right] \tag{7.40}$$

with the requirement

$$v \le \frac{\pi}{4}\left(\frac{w_0}{p}\right)^2 \tag{7.41}$$

where Eq. (4.18) has been used.

For uniform illumination, Eq. (7.38) becomes

$$u(z_v) = \frac{2vp^2 n_0 g_0 d_1}{d_1 \lambda - 2vp^2} \tag{7.42}$$

Likewise, Eq. (7.42) reduces to

$$u(z_v) = \frac{2vp^2 n_0 g_0}{\lambda} \tag{7.43}$$

for uniform plane illumination.

Finally, in a transverse quadratic medium for which g(z) is a constant g_0, the axial location of self–images for uniform plane illumination is given by [7.20]

$$z_v = \frac{1}{g_0}\tan^{-1}\left(\frac{2vp^2 n_0 g_0}{\lambda}\right) \tag{7.44}$$

We apply the results to a GRIN medium with a divergent linear taper function given by Eq. (4.98). For this case the axial location of self–images becomes

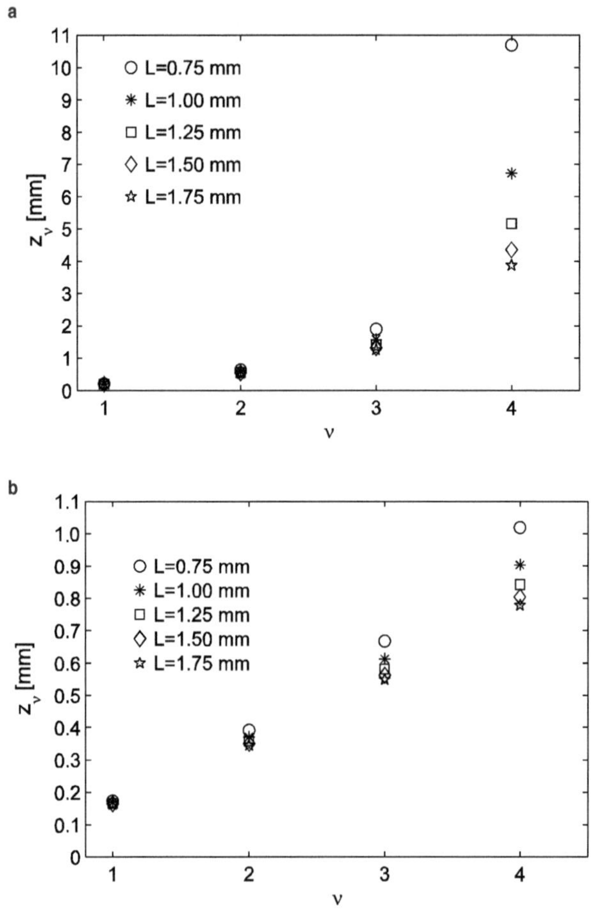

Fig. 7.3. Dependence of the self–image distances on L for **a** non–uniform and **b** uniform illumination. Calculations have been made for λ=0.7 μm, p=6 μm, $d_1 = 10$ mm, R(0)=0.65 mm, and w_0=7 μm.

$$z_\nu = L\left\{\exp\left[\frac{1}{g_0 L}\tan^{-1}\left[\frac{n_0 g_0 R^2(0) z_R^2 \lambda}{4\nu p^2\left[z_R^2 + n_0^2 R^2(0)\right] - 2\lambda R(0) z_R^2}\left(1 - \frac{4\nu p^2}{R(0)\lambda} - \sqrt{1 - \left(\frac{4 n_0 \nu p^2}{z_R \lambda}\right)^2}\right)\right]\right] - 1\right\} \tag{7.45}$$

for non–uniform illumination and

$$z_\nu = L\left\{\exp\left(\frac{1}{g_0 L}\tan^{-1}\left(\frac{2\nu p^2 n_0 g_0 d_1}{\lambda d_1 - 2\nu p^2}\right)\right) - 1\right\} \tag{7.46}$$

for uniform illumination, where Eqs. (7.38) and (7.42) have been used.

Equations (7.45–7.46) represent the discrete axial distances for which the phase coincidence between the interfering diffraction orders occurs.

We will analyze the dependence of the self–image distances on the taper function, the illumination, and the period of the object in this kind of GRIN medium.

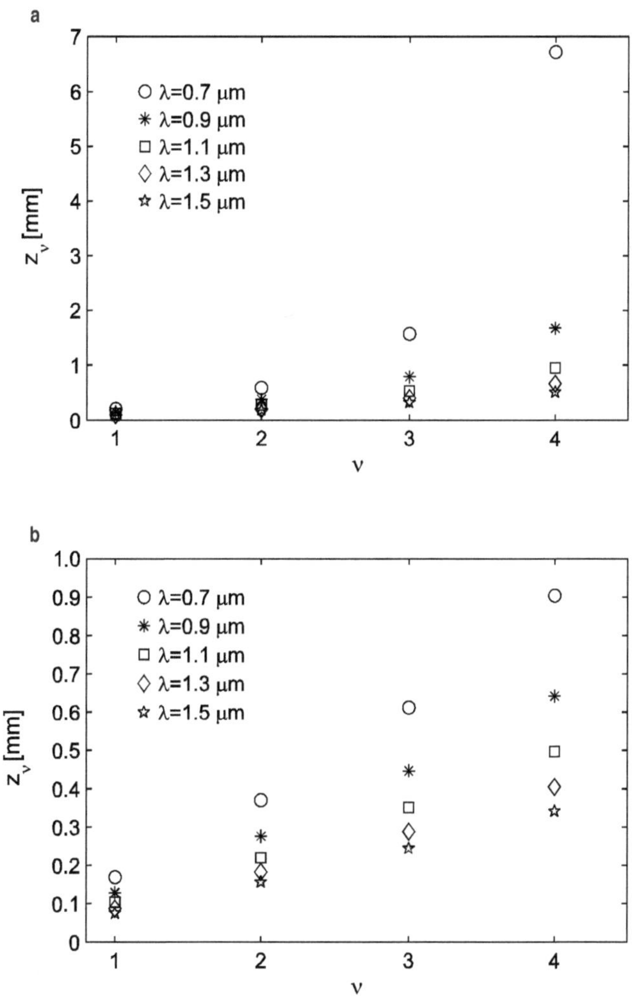

Fig. 7.4. Dependence of the self–image distances on λ for **a** non–uniform and **b** uniform illumination. Calculations have been made for p=6 µm, L=1 mm, $d_1 = 10\,\text{mm}$, R(0)=0.65 mm, and w_0=7 µm.

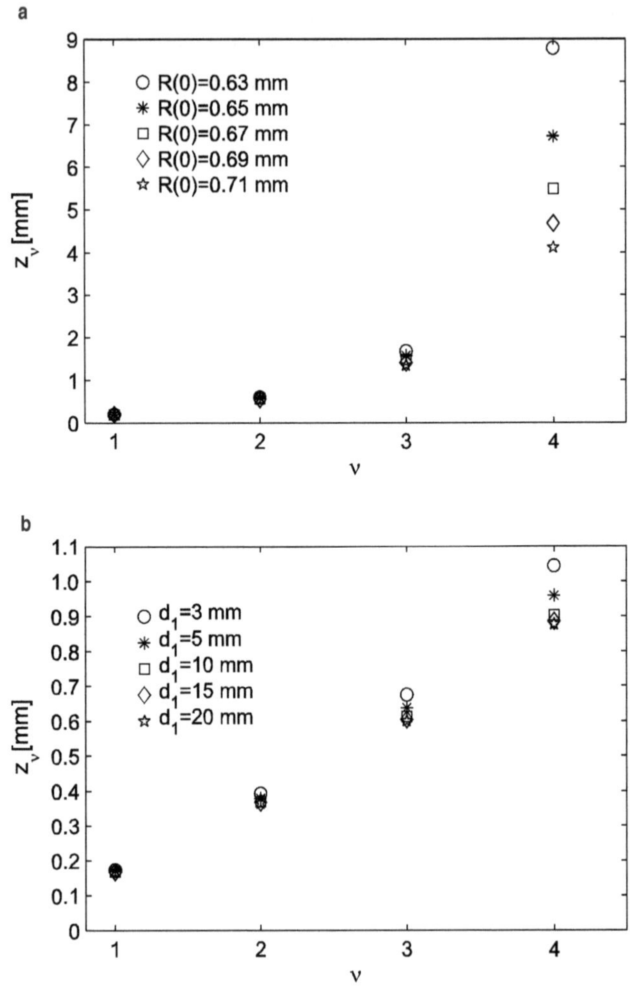

Fig. 7.5. Dependence of the self–image distances on the curvature radius R(0) and d_1 for **a** non–uniform and **b** uniform illumination, respectively. Calculations have been made for λ=0.7 µm, L=1 mm, p=6 µm, and w_0= 7µm.

Figures 7.3–7.6 show the axial locations of self–images versus self–image number for different values of taper parameters, wavelength, radius of curvature of illumination, and period of the object.

Figure 7.3 shows dependence on L for non–uniform illumination and uniform illumination. For a given L the interval between consecutive self–images increases with ν, and, in contrast, self–image locations decrease with L for a given ν. A similar behavior is shown in Figs. 7.4 and 7.5 for the dependence of the self–image distances on the wavelength and the curvature radius of the beam for non–

uniform illumination (a) and uniform illumination (b), respectively. Figure 7.6 represents the self–image evolution for several values of the object period. For a given p, the interval between consecutive self–images increases with ν, and, likewise, self–image locations increase with p for a given ν.

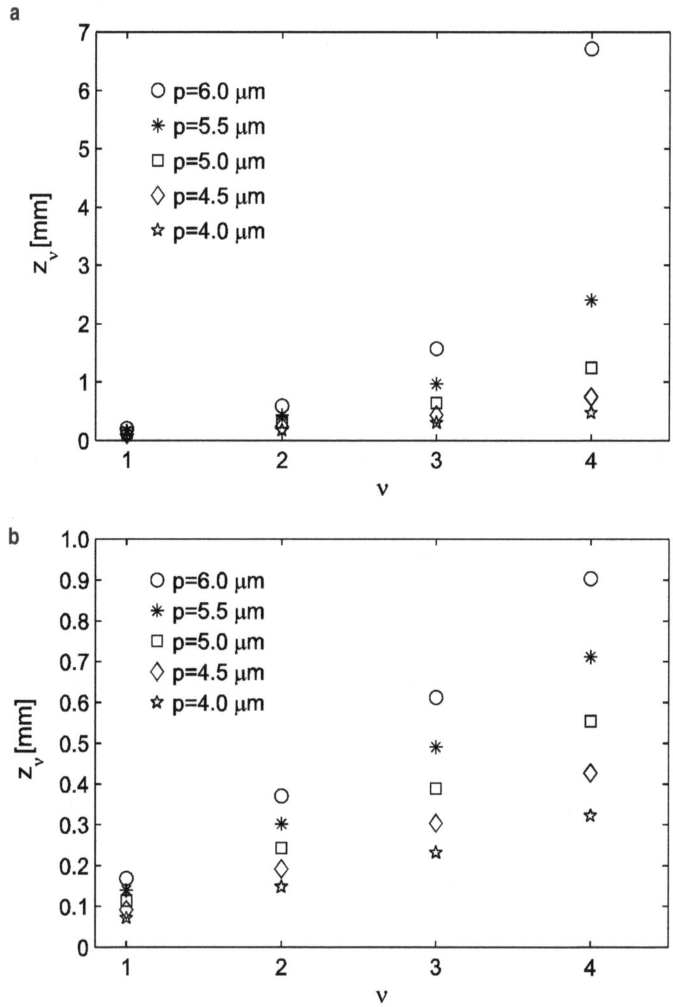

Fig. 7.6. Dependence of the self–image distances on p for **a** non–uniform and **b** uniform illumination. Calculations have been made for λ=0.7 μm, L=1 mm, d_1=10 mm, R(0)=0.65 mm, and w_0=7 μm.

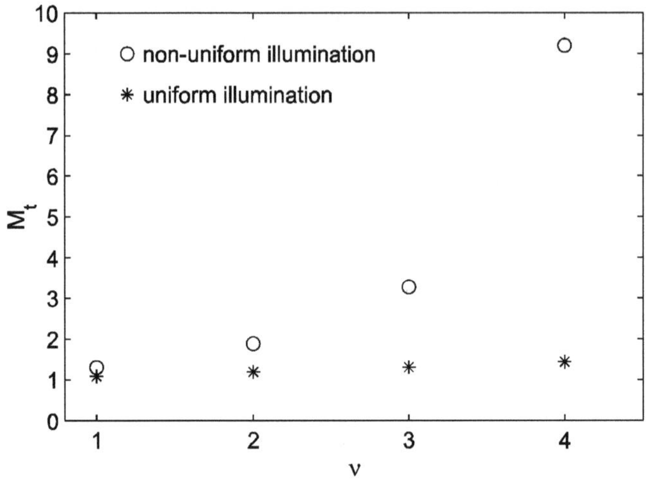

Fig. 7.7. Transverse magnification versus self–image number for non–uniform and uniform illumination. Calculations have been made for n_0=1.5, g_0=0.01 mm^{-1}, λ=0.7 µm, L=1 mm, $p = 6\,\mu m$, d_1=10 mm, R(0)=0.65 mm, and w_0=7 µm.

Figure 7.7 depicts the variation of transverse magnification with self–image number for non–uniform and uniform illumination. The first four self–image positions are obtained at z_1=0.202, z_2=0.584, z_3=1.549, and z_4=6.365 mm, for Gaussian illumination and at z_1=0.169, z_2=0.370, z_3=0.612 and z_4=0.903 mm for uniform illumination. A very slow dependence on self–image number is obtained for uniform illumination. On the other hand, for Gaussian illumination it is possible to obtain a pseudo–lens yielding images with significant lateral magnification of periodic objects for an object–image distance of about 6 mm [7.21–7.22].

Figures 7.3–7.7 indicate that the dependence of self–image location on taper profile, wavelength, curvature radius, and period object, as well as the variation of transverse magnification with self–image number for non–uniform illumination, is greater than that for uniform illumination. In all cases, calculations have been made for n_0=1.5 and g_0=0.01 mm^{-1}.

It is easy to prove that the reverse behavior occurs in a convergent linear GRIN medium whose taper function is given by

$$g(z) = \frac{g_0}{1 - \frac{z}{L}} \tag{7.47}$$

7.5 Fractional Talbot Effect: Unit Cell

The periodic object given by Eq. (7.1) also exhibits self–images at fractions of the Talbot distances [7.7, 7.23–7.26], that is, the Talbot condition now becomes

$$\frac{\lambda \operatorname{Re}[G(z_{\beta/\alpha})]}{n_0 p^2 |G(z_{\beta/\alpha})|^2} H_a(z_{\beta/\alpha}) = \frac{\beta}{\alpha} \tag{7.48}$$

for non–uniform illumination, and

$$\frac{\lambda H_a(z_{\beta/\alpha})}{n_0 p^2 F(z_{\beta/\alpha})} = \frac{\beta}{\alpha} \tag{7.49}$$

for uniform illumination, where α and β are integers. In particular, the fractional Talbot effect is obtained for the case where α and β are co–prime integers.

The fractional Talbot effect can also be considered the equivalent effect of an apparent lens of multifocal distance and transverse magnification given, respectively, by

$$f_{\beta/\alpha} = \frac{\lambda d_0^2 \alpha}{p^2 \beta} \tag{7.50}$$

$$M_t = \frac{|G(z_{\beta/\alpha})|^2}{\operatorname{Re}[G(z_{\beta/\alpha})]} \tag{7.51}$$

for non–uniform illumination, and

$$f^u_{\beta/\alpha} = \frac{\lambda d_{u0}^2 \alpha}{p^2 \beta} \tag{7.52}$$

$$M^u_t(z_{\beta/\alpha}) = F(z_{\beta/\alpha}) \tag{7.53}$$

for uniform illumination.

The axial localization of fractional self–images can be obtained from the fractional Talbot condition and, for instance, for uniform illumination is expressed as

$$\int_0^{z_{\beta/\alpha}} g(z')dz' = \tan^{-1}\left(\frac{n_0 g_0 d_1 p^2 \beta}{\lambda d_1 \alpha - p^2 \beta} \right) \tag{7.54}$$

where Eqs. (7.49), (3.92), and (1.92) have been used.

Figure 7.8 shows the axial position of self–images versus self–image number for a GRIN medium with a divergent linear taper function. This axial position is given by

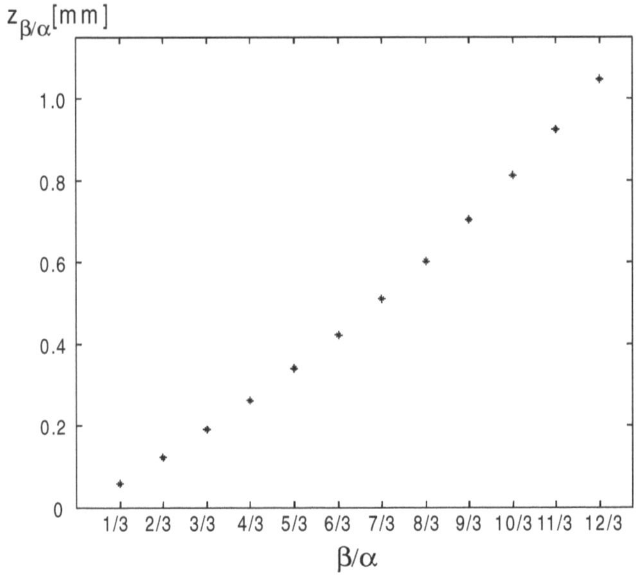

Fig. 7.8. Axial position of self–images versus self–image number. Calculations have been made for n_0=1.5, g_0=0.01 mm^{-1}, p=9 μm, λ=0.7 μm, L=1 mm, d_1=15 mm, and α=3.

$$z_{\beta/\alpha} = L\left\{\exp\left[\frac{1}{g_0}\tan^{-1}\left(\frac{n_0 g_0 d_1 p^2\beta}{d_1\alpha\lambda - p^2\beta}\right)\right] - 1\right\} \tag{7.55}$$

Figure 7.8 indicates that the interval between consecutive fractional self–images increases with β/α. For β/α= 3/3, 6/3, 9/3 and 12/3, the first four integer self–images are obtained.

To explain the effect of these results on the fractional Talbot effect, we consider an input object of period p, consisting of a Dirac comb, given by

$$T_c(x_0) = \sum_{q=-\infty}^{+\infty}\delta(x_0 - qp) = \frac{1}{p}\sum_{q=-\infty}^{+\infty}\exp\left\{-i\frac{2\pi q x_0}{p}\right\} \tag{7.56}$$

For the sake of simplicity, we suppose an incident uniform beam due to a cylindrical wavefront of radius of curvature d_1 such that

$$\psi_0(x_0) = \frac{1}{\sqrt{d_1}}\exp\left(i\frac{\pi x_0^2}{\lambda d_1}\right) \tag{7.57}$$

is the complex amplitude distribution at the input.

Inserting Eq. (7.57) into Eq. (7.5), taking into account Eq. (7.2), and integrating, we find that the complex amplitude distribution at fractional Talbot distances is written as

$$\phi_c(x;z_{\beta/\alpha}) = \frac{1}{\sqrt{d_1 F(z_{\beta/\alpha})}} \exp\left[ikn_0 z_{\beta/\alpha}\right] \exp\left[i\frac{kn_0\dot{F}(z_{\beta/\alpha})}{2F(z_{\beta/\alpha})}x^2\right]$$
$$\frac{1}{p}\sum_{q=-\infty}^{+\infty} \exp\left[-i\frac{2\pi qx}{pF(z_{\beta/\alpha})}\right] \exp\left[-i\frac{\pi q^2\beta}{\alpha}\right] \tag{7.58}$$

where Eq. (3.92) has been used.

After calculation, Eq. (7.58) can be rewritten as [7.27]

$$\phi_C(x;z_{\beta/\alpha}) = \left[\frac{F(z_{\beta/\alpha})}{\alpha d_1}\right]^{1/2} \exp\left[ikn_0 z_{\beta/\alpha}\right] \exp\left[i\frac{kn_0\dot{F}(z_{\beta/\alpha})}{2F(z_{\beta/\alpha})}x^2\right]$$
$$\sum_{q=-\infty}^{+\infty} \delta\left[x' - \frac{qpF(z_{\beta/\alpha})}{\alpha}\right] E(q;\alpha,\beta) \tag{7.59}$$

where

$$x' = x + \frac{1}{2}pF(z_{\beta/\alpha})e_\beta \tag{7.60}$$

with $e_\beta = 0(1)$ if β is even(odd), and E denoting the pure phase factor given by

$$E(q;\alpha,\beta) = \frac{1}{\sqrt{\alpha}}\sum_{s=1}^{\alpha} \exp\left\{i\frac{\pi}{\alpha}\left[2s\left(q + \frac{\alpha e_\beta}{2}\right) - \beta s^2\right]\right\} \tag{7.61}$$

From Eq. (7.59) it follows that the summation at $z_{\beta/\alpha}$ may be regarded as a replica of the input Dirac comb, with spacing $pF(z_{\beta/\alpha})/\alpha$, weighted by the phase factor E, since

$$\sum_{q=-\infty}^{+\infty} \delta\left[x' + pF(z_{\beta/\alpha}) - \frac{qpF(z_{\beta/\alpha})}{\alpha}\right] E(q;\alpha,\beta) = \sum_{r=-\infty}^{+\infty} \delta\left(x' - \frac{rpF(z_{\beta/\alpha})}{\alpha}\right) E(r;\alpha,\beta) \tag{7.62}$$

Then the fractional Talbot image reproduces periodically a unit cell of period $pF(z_{\beta/\alpha})$ containing α weighted images of the unit cell of the original Dirac comb and irradiance proportional to $|F(z_{\beta/\alpha})|/\alpha d_1$. If β is even, the Talbot image is centered at x=0, and if β is odd, the Talbot image is laterally shifted and centered at $x = -\frac{pF(z_{\beta/\alpha})}{2}$.

Next, we can return to the general periodic object given by Eq. (7.1), which is also expressed as

$$T(x_0) = \int_{-\infty}^{+\infty} t(\eta)\sum_{q=-\infty}^{+\infty}\delta(\eta - x_0 + qp)d\eta = \sum_{q=-\infty}^{+\infty} t(x_0 - qp) \tag{7.63}$$

where t is the transmittance function in the unit cell of the periodic object.

Substituting Eq. (7.63) into Eq. (7.5), the complex amplitude distribution at z>0 is given by

$$\psi(x;z)=\int_{-\infty}^{+\infty}t(\eta)\exp\left[ikn_0\dot{F}(z)\eta\left(x-\frac{\eta F(z)}{2}\right)\right]\psi_c(x-\eta F(z);z)d\eta \tag{7.64}$$

where

$$\psi_c(x;z)=\int_{-\infty}^{+\infty}K(x,x_0;z)\psi_0(x_0)\sum_{q=-\infty}^{+\infty}\delta(x_0-qp)dx_0 \tag{7.65}$$

Equation (7.64) at fractional Talbot distances reduces to

$$\begin{aligned}\psi(x;z_{\beta/\alpha})=&\left[\frac{1}{\alpha\, d_1F(z_{\beta/\alpha})}\right]^{1/2}\exp[ikn_0z_{\beta/\alpha}]\exp\left[i\frac{kn_0\dot{F}(z_{\beta/\alpha})}{2F(z_{\beta/\alpha})}x^2\right]\\&\sum_{q=-\infty}^{+\infty}t\left\{\frac{1}{F(z_{\beta/\alpha})}\left[x'-\frac{qpF(z_{\beta/\alpha})}{\alpha}\right]\right\}E(q;\alpha,\beta)\end{aligned} \tag{7.66}$$

where Eq. (7.59) has been used.

Equation (7.66) is the heart of the fractional Talbot effect. From a comparison of this equation and Eq. (7.59), it follows that at $z_{\beta/\alpha}$ each unit cell of a Talbot image of irradiance $1/\alpha d_1|F(z_{\beta/\alpha})|$ consists of α weighted images of the unit cell of the original periodic object with scaling factor $pF(z_{\beta/\alpha})/\alpha$. When β is even, the fractional Talbot image of period $pF(z_{\beta/\alpha})$ is centered with respect to the object, and when β is odd, the image presents a transverse shift. Likewise, the period of fractional self–images carries information about the transverse magnification of the periodic object due to uniform illumination given by $F(z_{\beta/\alpha})$, which coincides with Eq. (7.53).

Note that for $\alpha=1$, where $\beta/\alpha=\nu$ for integer values of ν, and $E(q;\alpha,\beta)=1$, Eq. (7.66) becomes

$$\psi(x,z_\nu)=\left[\frac{1}{d_1F(z_\nu)}\right]^{1/2}\exp[ikn_0z_\nu]\exp\left[i\frac{kn_0\dot{F}(z_\nu)}{2F(z_\nu)}x^2\right]\sum_{q=-\infty}^{+\infty}a_q\exp\left[-i\frac{2\pi qx'}{pF(z_\nu)}\right] \tag{7.67}$$

and integers Talbot images are obtained.

Finally, the irradiance distribution at the Talbot images is given by

$$\begin{aligned}I(x;z_{\beta/\alpha})=\left|\psi(x;z_{\beta/\alpha})\right|^2=&\frac{1}{d_1F(z_{\beta/\alpha})}\sum_{q,r=-\infty}^{\infty}a_qa_r^*\exp\left\{-i\frac{2\pi(q-r)x}{pF(z_{\beta/\alpha})}\right\}\\&\cdot\exp\left\{-i\frac{\pi\beta(q^2-r^2)}{\alpha}\right\}\end{aligned} \tag{7.68}$$

where Eq. (7.58) has been used.

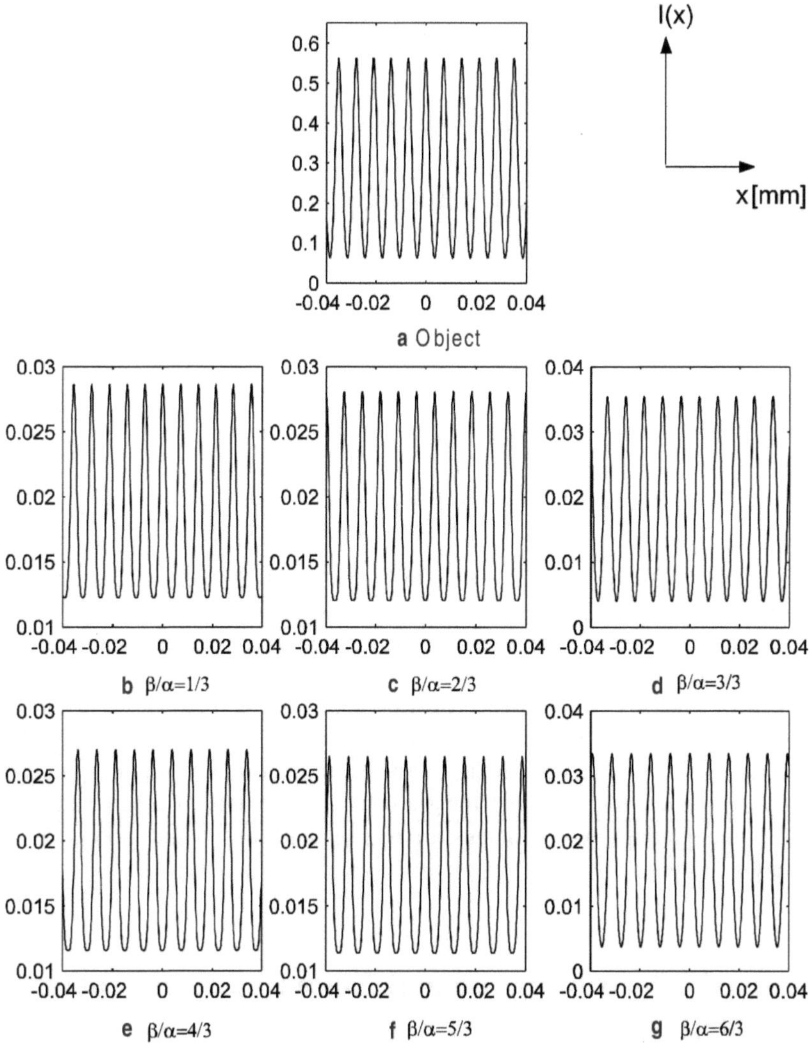

Fig. 7.9. Irradiance distribution at **a** the object, **b–c** and **e–f** some fractional Talbot images, and **d** and **g** the first two integer Talbot images. Calculations have been made for grating parameters A=1/2, B=1/8, and p=7 μm; GRIN parameters n_0=1.5, g_0=0.01 mm^{-1}, and L=1 mm; illumination parameters d_1=15 mm and λ =0.7 μm; and α =3. For β/α =3/3 and 6/3, the first two integer self–images are obtained.

In order to analyze the irradiance distribution at the integer and fractional Talbot images, we consider a hybrid system formed by a GRIN medium with a divergent linear taper function and a sinusoidal grating whose transmittance function is expressed as

$$T(x_0) = A + 2B\cos\left(\frac{2\pi x_0}{p}\right) = A + B\left[\exp\left\{i\frac{2\pi x_0}{p}\right\} + \exp\left\{-i\frac{2\pi x_0}{p}\right\}\right] \quad (7.69)$$

where A and B are parameters of modulation.

In this case, one finds for the irradiance at fractional Talbot images that [7.28].

$$I(x; z_{\beta/\alpha}) = \frac{2}{d_1 F(z_{\beta/\alpha})}\left\{\frac{A^2}{2} + 2B^2\cos^2\left[\frac{2\pi x}{pF(z_{\beta/\alpha})}\right] + 2AB\cos\left[\frac{2\pi x}{pF(z_{\beta/\alpha})}\right]\cos\left(\frac{\pi\beta}{\alpha}\right)\right\} \quad (7.70)$$

The above equation, for integer Talbot distances, becomes

$$I(x; z_v) = \frac{2}{d_1 F(z_v)}\left\{\frac{A}{\sqrt{2}} + (-1)^v\sqrt{2}B\cos\left[\frac{2\pi x}{pF(z_v)}\right]\right\}^2 \quad (7.71)$$

From Eqs. (7.70) or (7.71) it follows that at Talbot distances, irradiance patterns with period pF are observed. Each integer or fractional Talbot image arises from interference between diffraction orders of the sinusoidal grating through the GRIN medium.

Figure 7.9 shows the irradiance distribution at (a) the grating; (b, c and e, f) some fractional Talbot images; and (d and g) the first two integer Talbot images. The maximum contrast, equal to the grating contrast, coincides with integer Talbot images. However, for fractional images contrast decreases to a half value from that in the grating. A continuous transverse shift of images occurs as β/α increases.

7.6 Effect of Off–Axis Source and Finite Object Dimension on Self–Images

Firstly, we suppose an off–axis source located at a point ζ in transverse direction from axis. The complex distribution on the periodic object is now given by

$$\psi(x_0) = \frac{1}{\sqrt{d_1}}\exp\left\{i\frac{\pi(x_0 - \zeta)^2}{\lambda d_1}\right\}T(x_0) \quad (7.72)$$

Substitution of Eq. (7.72) into Eq. (7.5) and integration provides

$$\psi(x,z)=\left[\frac{1}{d_1F(z)}\right]^{1/2}\exp\{ikn_0z\}\exp\left\{i\frac{kn_0\dot{F}(z)}{2F(z)}x^2\right\}\exp\left\{-i\frac{k\zeta x}{d_1F(z)}\right\}$$
$$\cdot\sum_{q=-\infty}^{+\infty}a_q\exp\left\{-i\frac{2\pi q}{pF(z)}\left(x+\frac{H_a(z)\zeta}{n_0d_1}\right)\right\}\exp\left\{-i\frac{\pi\lambda H_a(z)q^2}{n_0p^2F(z)}\right\} \quad (7.73)$$

From Eq. (7.73), it follows that the irradiance at Talbot distances can be written as

$$I(x;z_{\beta/\alpha})=\frac{1}{d_1F(z_{\beta/\alpha})}\sum_{q,r=-\infty}^{\infty}a_qa_r^*\exp\left\{-i\frac{2\pi(q-r)}{pF(z_{\beta/\alpha})}\left(x+\frac{H_a(z_{\beta/\alpha})\zeta}{n_0d_1}\right)\right\}$$
$$\cdot\exp\left\{-i\frac{\pi\beta(q^2-r^2)}{\alpha}\right\} \quad (7.74)$$

If we consider above hybrid system, in order to derive the effects arising as a result of displacement of the source, Eq. (7.74) becomes

$$I(x;z_{\beta/\alpha})=\frac{2}{d_1F(z_{\beta/\alpha})}\left\{\frac{A^2}{2}+2B^2\cos^2\left[\frac{2\pi}{pF(z_{\beta/\alpha})}\left(x+\frac{H_a(z_{\beta/\alpha})\zeta}{n_0d_1}\right)\right]\right.$$
$$\left.+2AB\cos\left[\frac{2\pi}{pF(z_{\beta/\alpha})}\left(x+\frac{H_a(z_{\beta/\alpha})\zeta}{n_0d_1}\right)\right]\cos\left[\frac{\pi\beta}{\alpha}\right]\right\} \quad (7.75)$$

at fractional Talbot images and

$$I(x;z_v)=\frac{2}{d_1F(z_v)}\left\{\frac{A}{\sqrt{2}}+(-1)^v\sqrt{2}B\cos\left[\frac{2\pi}{pF(z_v)}\left(x+\frac{H_a(z_v)\zeta}{n_0d_1}\right)\right]\right\} \quad (7.76)$$

at integer Talbot images.

From Eqs. (7.75–76) and Eqs. (7.70–71) it follows that an example of the difference between on–axis and off–axis illuminations concerns the lateral shift of irradiance pattern [7.29], since this pattern is centered along a geometrical trajectory proportional to the axial ray which is given by

$$x(z)=-\dot{x}_0H_a(z) \quad (7.77)$$

where $\dot{x}_0=\zeta/n_0d_1$ is the slope of the refracted ray at the input.

Figure 7.10 depicts for comparison, the irradiance distribution at the two first fractional images and the first integer image of figure 7.9 for off–axis (dot line) and on–axis(solid line) illuminations. A lateral shift in reverse sense between both patterns is observed for $\zeta=2\text{mm}$. On the contrary, lateral shift in same sense would be obtained for $\zeta=-2\text{mm}$.

After that, we consider a hybrid system of finite dimension with aperture 2a illuminated by an on–axis source. In this case, the complex amplitude distribution in the GRIN medium at $z>0$ becomes

$$\psi(x;z) = \int_{-a}^{+a} \psi(x_0) K(x, x_0; z) dx_0 \tag{7.78}$$

Inserting Eqs. (7.1–7.2) and (7.57) into Eq. (7.78) and integrating, we arrive at the following complex amplitude distribution

$$\begin{aligned}\psi(x;z) = &\left[\frac{1}{i2d_1F(z)}\right]^{1/2} \exp[ikn_0z]\exp\left[i\frac{kn_0\dot{F}(z)}{2F(z)}x^2\right] \\ &\sum_{q=-\infty}^{+\infty} a_q \exp\left\{-i\frac{2\pi qx}{pF(z)}\right\}\exp\left\{-i\frac{\pi\lambda H_a(z)q^2}{n_0p^2F(z)}\right\}\left[Q(\theta_q^+) - Q(\theta_q^-)\right]\end{aligned} \tag{7.79}$$

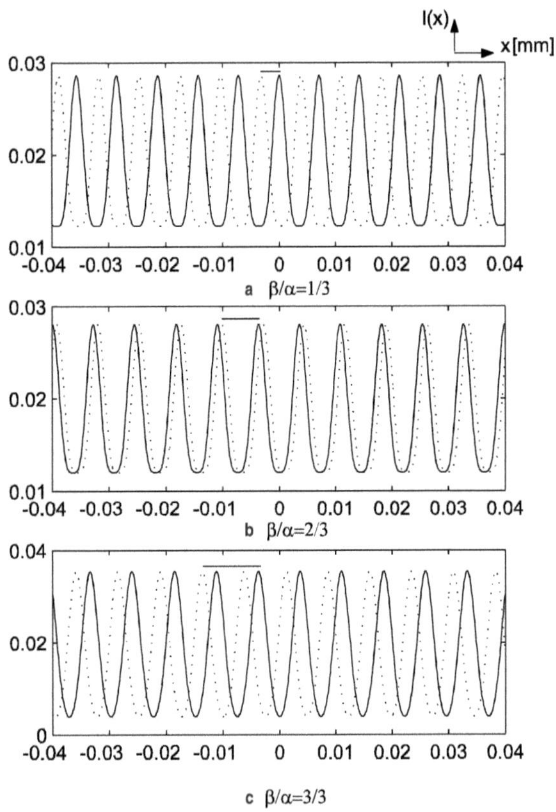

Fig. 7.10. Irradiance distribution for on–axis (solid line) and off–axis illumination (dotted line) at **a–b** the first two fractional, and **c** the first integer Talbot images. Calculations have been made for parameters of Fig. 7.9 and $\zeta = 2$mm.

where Q is the complex Fresnel integral defined as

$$Q(\tau) = \int_0^\tau \exp\left\{i\frac{\pi t^2}{2}\right\}dt \tag{7.80}$$

and

$$\theta_q^{\pm} = \sqrt{\frac{2n_0F(z)}{\lambda H_a(z)}}\left[\pm a - \frac{1}{F(z)}\left(x + \frac{\lambda q H_a(z)}{n_0 p}\right)\right] \tag{7.81}$$

For the Talbot condition, Eq. (7.79) gives

$$\psi(x;z_{\beta/\alpha}) = \left[\frac{1}{i2d_1F(z_{\beta/\alpha})}\right]^{1/2} \exp\left[ikn_0z_{\beta/\alpha}\right]\exp\left[i\frac{kn_0\dot{F}(z_{\beta/\alpha})}{2F(z_{\beta/\alpha})}x^2\right]$$
$$\sum_{q=-\infty}^{+\infty} a_q \exp\left\{-i\frac{2\pi qx}{pF(z_{\beta/\alpha})}\right\}\exp\left\{-i\frac{\pi\beta q^2}{\alpha}\right\}\left[Q(\vartheta_q^+) - Q(\vartheta_q^-)\right] \tag{7.82}$$

where

$$\vartheta_q^{\pm} = \sqrt{\frac{2\alpha}{\beta p^2}}\left[\pm a - \left(\frac{x}{F(z_{\beta/\alpha})} + \frac{q\beta p}{\alpha}\right)\right] \tag{7.83}$$

are the arguments of Q evaluated at the Talbot condition.

From Eq. (7.82) it follows that the irradiance at Talbot images can be expressed as

$$I(x;z_{\beta/\alpha}) = \frac{1}{2d_1F(z_{\beta/\alpha})}\sum_{q,r=-\infty}^{\infty} a_q a_r^* \exp\left\{-i\frac{2\pi(q-r)x}{pF(z_{\beta/\alpha})}\right\}\exp\left\{-i\frac{\pi\beta(q^2-r^2)}{\alpha}\right\}$$
$$\cdot[Q(\vartheta_q^+) - Q(\vartheta_q^-)][Q^*(\vartheta_r^+) - Q^*(\vartheta_r^-)] \tag{7.84}$$

Equation (7.84) applied to the hybrid system formed by the GRIN medium and the sinusoidal grating gives

$$\begin{aligned}
I(x;z_{\beta/\alpha}) = \frac{1}{2d_1F(z_{\beta/\alpha})}\Bigg\{ & A^2\left[\left|Q(\vartheta_0^+)\right|^2 + \left|Q(\vartheta_0^-)\right|^2 - 2\,\mathrm{Re}\left[Q(\vartheta_0^+)Q^*(\vartheta_0^-)\right]\right] \\
& + B^2\left[\left|Q(\vartheta_1^+)\right|^2 + \left|Q(\vartheta_1^-)\right|^2 + \left|Q(\vartheta_{-1}^+)\right|^2 + \left|Q(\vartheta_{-1}^-)\right|^2 - 2\,\mathrm{Re}\left[Q(\vartheta_1^+)Q^*(\vartheta_1^-)\right]\right. \\
& \left. - 2\,\mathrm{Re}\left[Q(\vartheta_{-1}^+)Q^*(\vartheta_{-1}^-)\right]\right] + 2AB\,\mathrm{Re}\Bigg[\exp\left(\frac{i\pi\beta}{\alpha}\right)\left[Q(\vartheta_0^+) - Q(\vartheta_0^-)\right] \\
& \cdot\Bigg(\exp\left\{i\frac{2\pi x}{pF(z_{\beta/\alpha})}\right\}\left[Q^*(\vartheta_1^+) - Q^*(\vartheta_1^-)\right] \\
& + \exp\left\{-i\frac{2\pi x}{pF(z_{\beta/\alpha})}\right\}\left[Q^*(\vartheta_{-1}^+) - Q^*(\vartheta_{-1}^-)\right]\Bigg)\Bigg]\Bigg\}
\end{aligned} \tag{7.85}$$

where 0, ±1 subscripts correspond to 0, ±1 harmonics of the grating, respectively, and Re denotes the real part.

Figure 7.11 depicts the effects on fractional and integer Talbot images of figure 7.9 which arise as a result of finite object dimension. At the center of the images, the quality of the gratings decreases with z from that in the case of an infinite object. Diffraction around the edge of the geometrical shadow (x=aF) of the grating is observed.

Likewise, because of finite object dimension, the ±1 harmonics are gradually and laterally displaced from the zero–order harmonic, called the "walk–off" effect [7.30], and do not take part in the image formation when using an object of only a few unit cells or repeated elements as shown in Fig. 7.12. The irradiance pattern at the first integer Talbot image is represented for aperture sizes $2a = 3p$ (a), $2a = 5p$ (b), $2a = 10p$ (c), and $2a = 20p$ (d) where p is the spatial period of the grating. We can discuss image formation from case (c), for which the aperture is an order of magnitude higher than the spatial period.

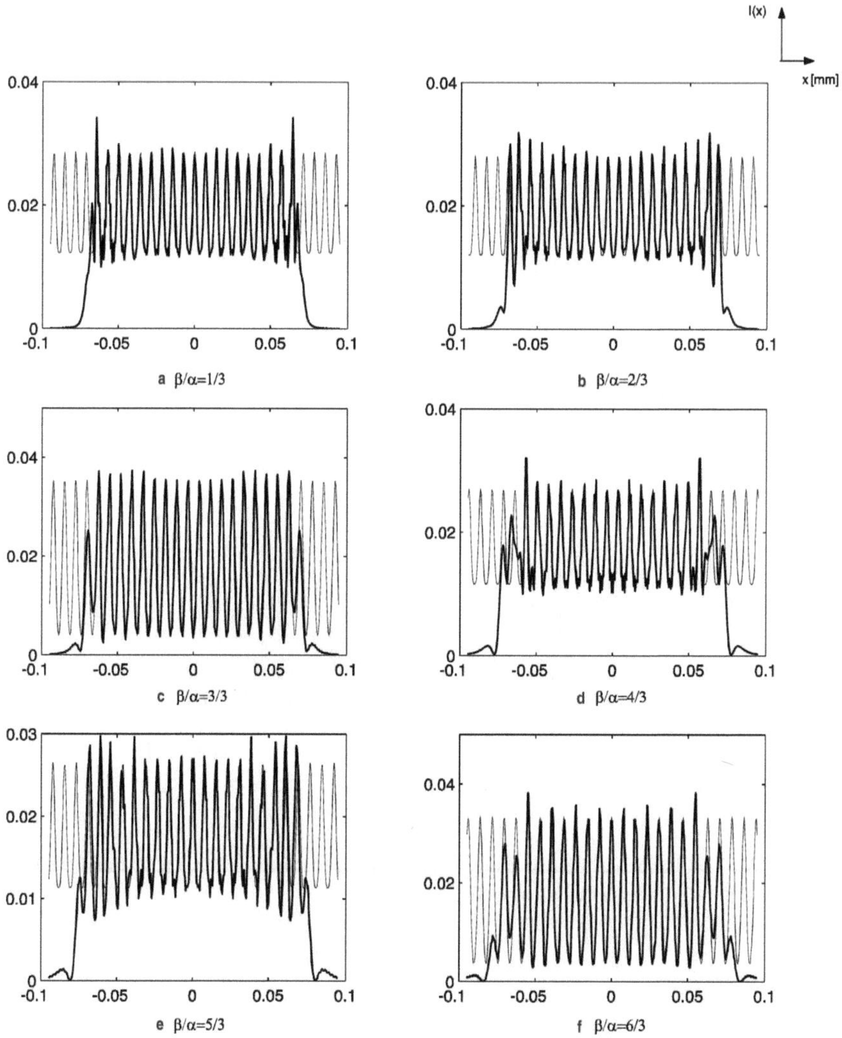

Fig. 7.11. Irradiance distribution at the cases depicted in Fig. 7.9. Calculations have been made for parameters of Fig. 7.9 and aperture 140 μm. The edge of the geometrical shadow is obtained for **a** x=71.35 μm, **b** x=72.73 μm, **c** x=74.13 μm, **d** x=75.58 μm, **e** x=77.05 μm, and **f** x=78.56μm. The thicker curves denote those without the effects of the aperture size.

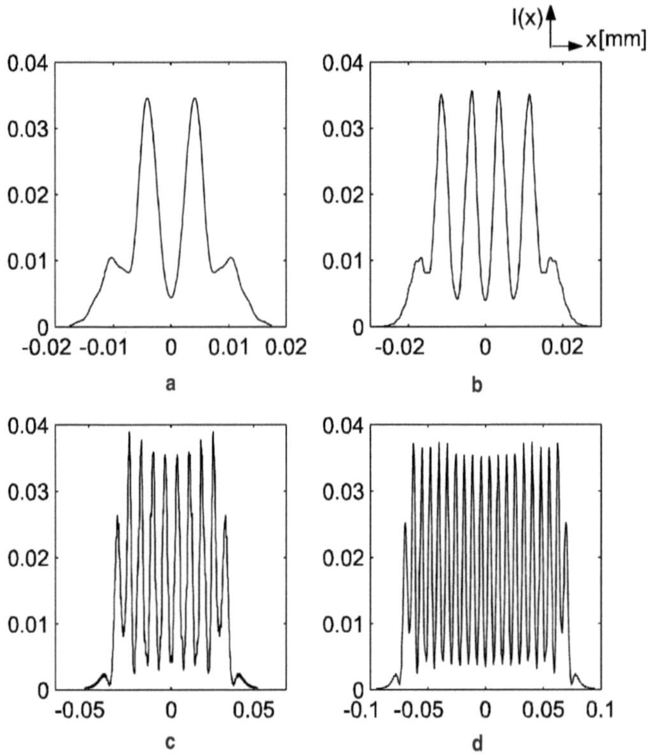

Fig. 7.12. Walk–off effect at the first integer Talbot effect for aperture sizes **a** 21 μm, **b** 35 μm, **c** 70 μm, and **d** 140 μm where in all cases p=7 μm.

8 GRIN Crystalline Lens

8.1 Introduction

The eye is the main organ for sensing light, and its design is optimized for capturing light and forming images. The eye is, in many respects, like a camera. The optical system of the eye forms an image on the retina, whereas the camera lens forms its image on film. The camera must be focused by changing the distance from the lens to the film. However, a change in the configuration of the crystalline lens (human lens) occurs when the eye needs to focus at different distances (accommodation). This involves alterations in curvature, thickness, and refractive index of the lens. A change in the axial length of the eye is not involved.

On the other hand, the refractive index of the human lens is not constant. The index value is almost constant in the nuclear region but decreases in the cortex of the lens. In the optical modeling of the crystalline lens, two different models are used. These are the continuous gradient index and the shell models. For the continuous gradient index model, the refractive index distribution is represented by continuous isoindicial surfaces. In the shell model, the gradient index is represented by a finite and discrete set of concentric shells, with the refractive index constant in each shell.

In this chapter we study light propagation through the lens by paraxial rays, assuming the continuous gradient index model. The axial and field rays and the gradient parameter as well as the refractive power and positions of the cardinal points of the lens will be obtained. A deeper knowledge of light propagation in the crystalline lens within the framework of GRIN optics can be an important goal for investigation in optometry and vision sciences.

8.2 The Optical Structure of the Human Eye

The structure of the human eye is shown in Fig. 8.1. The human eye is very nearly spherical in shape whose outer layer is the sclera. This dense, opaque, white layer is a membrane made of fibrous tissue, mainly protective in function. The front part of the sclera, and continuous with it, is a transparent section called the cornea.

The next layer of the eye is the uveal tract comprising the iris, the ciliary body and the choroid. The iris forms the aperture stop of the eye; at its center is an opening known as the pupil that automatically controls not only the amount of light falling onto the retina, but also the amount of light returning out of the eye. This process is known as adaptation. The ciliary body plays an important role in the accomodation process. Accomodation is the ability of the eye to change its power to focus on objects at different distances. The crystalline lens is attached to the ciliary body via the zonules. When relaxed, the eye is focused on objects at infinity. When it is desired to view an object nearer than infinity, contraction of the ciliary muscle leads to changes in zonular tension, which alters lens shape, causing a variation in the refractive power of the lens, and hence in the whole power of the eye. The choroid supports important vegetative processes.

The inner layer of the eye is the retin,a which is an extension of the central nervous system. The retina is connected to the brain by the optic nerve. The retina contains millions of cells, most 1–5μm in diameter at the cell body. There are six main types of neural cells in the retina, and of these, the photoreceptor cells (rods and cones) are responsible for capturing light. The cones predominate in the fovea, which is approximately 5° wide. The fovea is free of rods in its central 1° field. Rods reach their maximum density at about 20° from the fovea. The vascular supply to the retina enters the eye at the optic disc. There are no cones or rods here, and hence this region is blind. The name given to the corresponding region in the visual field is the blind spot [8.1].

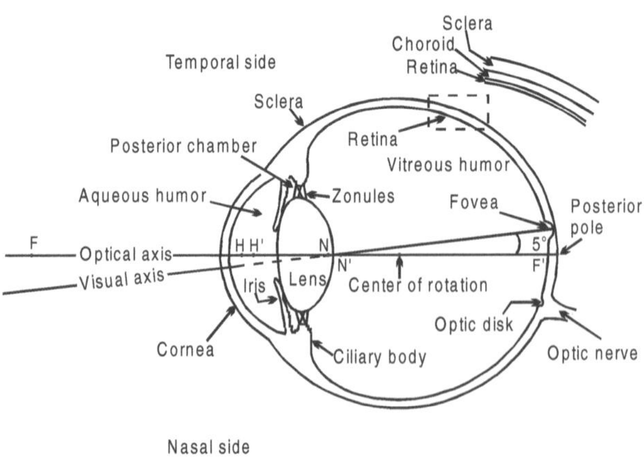

Fig. 8.1. Horizontal section of the right eye. Cardinal points are for the relaxed eye.

The inside of the eye is divided in three compartiments: the anterior chamber, between the cornea and iris, which contains a salt solution known as the aqueous humor; the posterior chamber, between the iris, the ciliary body, and the lens, which also contains the aqueous humor and; the vitreous chamber, between the lens and the retina, which contains a transparent gelatinous substance called vitreous humor.

The refracting elements of the eye are the cornea and the lens. Refraction takes place at four surfaces: the anterior and posterior interfaces of the cornea and the lens. There is also continuous refraction within the lens. The eye has a number of axes; two of these are the optical axis and the visual axis. The optical axis is defined as the line of best fit through the centers of curvature of the four refracting surfaces because of the lack of symmetry of the eye. The visual axis is the line joining the fixation point and the foveal image by way of the nodal points.

The cornea is the initial and major refracting element of the eye. It is aspheric. The shape of corneal surfaces, over a central zone of 8 mm of diameter, are often represents by conicoids expressed as [8.2]

$$r^2 = 2R_0 z - (1+Q)z^2 \tag{8.1}$$

where the origin is chosen at the vertex of the surface, r is the transverse distance from any point on the surface to the z–optical axis, R_0 is the radius of curvature at the vertex, and Q is the asphericity parameter. Often asphericity is expressed in terms of a quantity ρ called the shape factor, which is related to Q by

$$\rho = 1 + Q \tag{8.2}$$

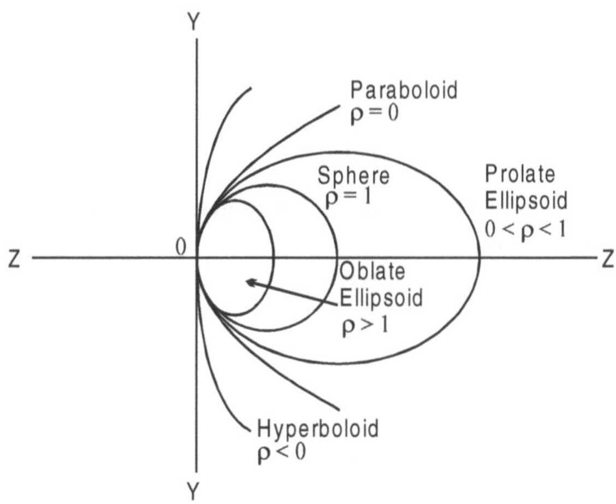

Fig. 8.2. Family of conic sections having the same apical radius but different shape factors.

The value and sign of ρ (or Q) specify the type of conicoid. Figure 8.2 shows a family of conic sections having the same radius of curvature at the vertex but different shape factors. Usual values of the vertex curvature radii of the anterior and posterior corneal surfaces are 7.7 mm and 6.5 mm, respectively. The cornea has an axial thickness of about 0.55 mm and a refractive index of 1.376.

The crystalline lens is a mass of cellular tissue of gradient index, contained within an elastic capsule attached to the ciliary body. The shape of the front and back surfaces can be also expressed by conicoids. Shape and internal form of the lens change with accommodation and age [8.3–8.8]. The most common values of the vertex radii of curvature of the unaccommodated or relaxed eye are +10 mm and –6 mm for the anterior and posterior surfaces of the lens, respectively. The lens thickness is often taken to be about 3.6 mm in the relaxed state, but the lens thickens upon accommodation and with increasing age. The equatorial diameter of the lens is between 8 mm and 10 mm, decreasing with accommodation. The refractive index within the lens is not constant and will be studied in the Sect. 8.3–8.5. The indices of refraction of both the aqueous humor and the vitreous humor are nearly equal to that of water, about 1.336. Since the refractive indices of the media that light rays traverse in the eye are all about the same order of magnitude, most of refraction occurs at the anterior surface of the cornea separating air and the eye.

Finally, any optical system has three pairs of cardinal points that lie on the optical axis, as discussed in Sect. 3.3. There are a number of useful equations connecting the cardinal points (Fig. 8.1), such as

$$P = -\frac{n_1}{HF} = \frac{n_1'}{H'F'} \tag{8.3}$$

$$HN = H'N' = \frac{n_1' - n_1}{P} \tag{8.4}$$

$$FN = H'F' \tag{8.5}$$

$$FH = N'F' \tag{8.6}$$

where n_1 and n_1' are the refractive indices of object space (air) and image space (vitreous humor) respectively.

Equation (8.3) gives the refractive power of the eye, the remain equations permit us to find the distance between principal and nodal points, Eq. (8.4), and the image and object focal lengths, Eqs. (8.5–8.6). Since the mean power of a relaxed eye is about 60 D (diopters), and the values of n_1 and n_1' are 1 and 1.336, we have $HN = H'N' = 5.6\,\text{mm}$, $FH=N'F'=16.7$ mm , and $FN=H'F'=22.3$ mm. Then, the object and image focal lengths of the eye are –16.7 mm and +22.3 mm, respectively.

8.3 The GRIN Model of the Crystalline Lens

The lens of the eye is a GRIN structure that changes its shape with accommodation [8.9–8.12]. Two different models are used in the optical modeling of the lens. These are the shell and the continuous gradient index models. The shell or laminated model is represented by a finite and discrete set of concentric shells, with the refractive index constant in each shell [8.13–8.15]. In the construction of such a model, it is necessary to decide the number of shells, how the refractive index varies from shell to shell, and the value of curvatures of the surface of each shell. Once the shell structure is established, paraxial ray–tracing can be used to determine the lens power [8.16]. For the continuous GRIN model the refractive index profile is represented by continuous isoindicial surfaces. Smith et al. [8.17] described four models pertaining to the external shape of the lens and the internal contours of refractive index. For the first two models the refractive index distribution is represented by bi–elliptical isoindicial surfaces that are concentric with the lens surfaces. The first model assumes symmetric isoindicial surfaces, and the second one is an asymmetric model in which the posterior curvature of any isoindicial surface is greather than that of the anterior in such a way that, at the equatorial plane of joining, the isoindicial surface is smooth and continuous as shown in Fig. 8.3. Such a model has also been proposed by Jagger to describe the cat lens [8.18]. The third model is a modified version of the second model, in which the isoindicial surfaces remain asymmetric ellipsoids, but the surfaces of the lens are no longer isoindicial contours. The last model is more general, allowing any conicoid surface shape and a non–smooth joint at the equator.

The second and third models provide a closest simulation to the real situation; we will describe the second model here. With this model, the front surface of the crystalline lens in the saggital section is described by the ellipse (Fig. 8.4a)

$$\frac{(z-a_1)^2}{a_1^2}+\frac{y^2}{b^2}=1 \tag{8.7}$$

where a_1 is the semiaxis along the z–optical axis, and b is the semiaxis along the y–axis, that is, the equatorial radius of the lens. For this surface the origin of the axes is at the anterior vertex.

The back surface is represented by

$$\frac{z^2}{a_2^2}+\frac{y^2}{b^2}=1 \tag{8.8}$$

where a_2 is the semiaxis along the z–axis, with the origin of the axes at the center of the equatorial section.

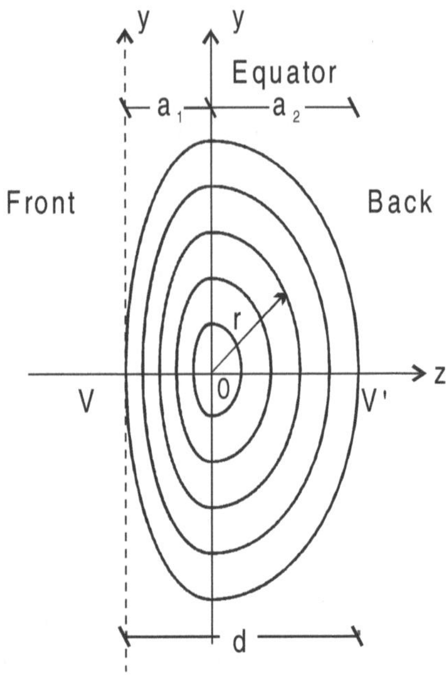

Fig. 8.3. Refractive index distributions of the lens of the eye in the sagittal section represented as bi–elliptical isoindicial curves joined at the equator. For the front surface the origin of the axes is at V, and for the back surface the origin is at 0.

From Eqs. (8.7–8.8) it follows that the curvature radii of the front and back surfaces at the vertices V and V′ are given, respectively, by

$$r_V = \frac{b^2}{a_1} \tag{8.9a}$$

$$r_{V'} = -\frac{b^2}{a_2} \tag{8.9b}$$

On the other hand, in any direction from the center to the edge of the lens, the refractive index profile is given by a polynomial of the form [8.19]

$$n(r) = \sum_{j=0}^{\infty} c_j r^{2j} = c_0 + c_1 r^2 + c_2 r^4 + \ldots \tag{8.10}$$

where c_0, c_1, etc., are coefficients, and r is the normalized distance in any direction with value of 1 at the edge. Equation (8.10) does not include any odd terms in r if we assume the lens is rotationally symmetric about the optical axis.

If the central index is n_c and the edge index is n_e, we have the following conditions

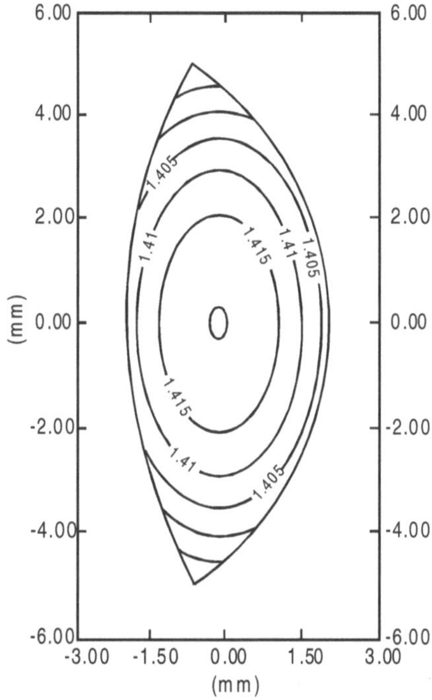

Fig. 8.4. Isoindicial curves in the sagittal section for the unaccommodated model of the crystalline lens proposed by Blaker [8.22].

$$n(0) = n_c = c_0 \tag{8.11a}$$

$$n(1) = n_e = \sum_{j=0}^{\infty} c_j \tag{8.11b}$$

and then

$$\Delta n = n_e - n_c = \sum_{j=1}^{\infty} c_j \tag{8.12}$$

where Δn is the difference in refractive index between the edge and center of the crystalline lens.

In the sagittal section, for the asymmetric bi–elliptical isoindicial surface model of the lens, the refractive index at any point in this section can be written as a power series

$$n(y,z) = \sum_{j=0}^{\infty} c_j f^j(y,z) \tag{8.13}$$

with

$$f(y,z)=\frac{(z-a_1)^2}{a_1^2}+\frac{y^2}{b^2} \tag{8.14a}$$

for the front part of the lens, and

$$f(y,z)=\frac{z^2}{a_2^2}+\frac{y^2}{b^2} \tag{8.14b}$$

for the back part of the lens.

At the equator, the refractive index profile and its derivatives with respect to y of odd and even orders are given by

$$n(y)=n_f(y,z)\big|_{z=a_1}=n_b(y,z)\big|_{z=0}=\sum_{j=0}^{\infty}c_j\left(\frac{y}{b}\right)^{2j} \tag{8.15a}$$

and

$$\frac{d^p n(y)}{dy^p}=\frac{1}{b^p}\sum_{j=q}^{\infty}2j(2j-1)...(2j-p+1)c_j\left(\frac{y}{b}\right)^{2j-p} \tag{8.15b}$$

where q=p/2 for p even, q=(p+1)/2 for p odd, and subscripts f and g denote front and back parts of the lens, respectively.

At the center of the equator the derivatives cancel, and at the edge they reduce to

$$\left.\frac{d^p n(y)}{dy^p}\right|_{y=b}=\frac{1}{b^p}\sum_{j=q}^{\infty}2j(2j-1)...(2j-p+1)c_j \tag{8.16}$$

Along the optical axis, the refractive index profile is given by

$$n_{0f}(z)=n_f(y,z)\big|_{y=0}=\sum_{j=0}^{\infty}c_j\left(\frac{z-a_1}{a_1}\right)^{2j} \tag{8.17a}$$

for the front part and

$$n_{0b}(z)=n_b(y,z)\big|_{y=0}=\sum_{j=0}^{\infty}c_j\left(\frac{z}{a_2}\right)^{2j} \tag{8.17b}$$

for the back part, such that $n_{0f}(a_1)=n_{0b}(0)=c_0$.

The derivatives with respect to z are written as

$$\frac{d^p n_{0f}(z)}{dz^p}=\frac{1}{a_1^p}\sum_{j=q}^{\infty}2j(2j-1)...(2j-p+1)c_j\left(\frac{z-a_1}{a_1}\right)^{2j-p} \tag{8.18a}$$

$$\frac{d^p n_{0b}(z)}{dz^p}=\frac{1}{a_2^p}\sum_{j=q}^{\infty}2j(2j-1)...(2j-p+1)c_j\left(\frac{z}{a_2}\right)^{2j-p} \tag{8.18b}$$

where q=p/2 for p even, and q=(p+1)/2 for p odd.

Note that at the center all derivatives vanish, and at the vertex they verify the condition

$$\frac{\left.\dfrac{d^p n_{0f}(z)}{dz^p}\right|_{z=0}}{\left.\dfrac{d^p n_{0b}(z)}{dz^p}\right|_{z=a_2}} = (-1)^p \left(\frac{a_2}{a_1}\right)^p \tag{8.19}$$

Likewise, the refractive index profile given by Eq. (8.13) can also be expressed as

$$n(y,z) = \sum_{j=0}^{\infty} n_j(z) y^{2j} \tag{8.20}$$

where $n_0(z)$, $n_1(z)$, etc, are polynomials in z given in compact form by [8.20–8.21]

$$n_j(z) = \sum_{i=0}^{\infty} n_{ji} z^i \tag{8.21}$$

Comparing Eq. (8.21) with Eq. (8.13) and taking into account Eqs. (8.14), the n_{ji} coefficients in terms of c_j coefficients can be evaluated for both parts of the lens.

If we neglect in the terms of higher order the second in Eq. (8.20), we have

$$n(y,z) = n_{00} + n_{01} z + n_{02} z^2 + n_{10} y^2 \tag{8.22}$$

Equation (8.22) describes a parabolic GRIN distribution corresponding to the model proposed by Blaker [8.22]. Figure 8.4 shows some isoindicial surfaces for an unaccommodated lens, and we can see that the refractive index decreases toward the edge of the crystalline lens [8.23].

Likewise, Gullstrand expressed the refractive index of the lens up to a fourth–order term in y as follows [8.24]

$$\begin{aligned} n(y,z) = {} & 1.406 - 0.0062685(z-z_0)^2 + 0.0003834(z-z_0)^3 \\ & - \left[0.00052375 + 0.00005735(z-z_0) + 0.00027875(z-z_0)^2\right] y^2 \\ & - 0.000066716 y^4 \end{aligned} \tag{8.23}$$

where $z_0 = 1.7$ mm is the position along the axis at which the refractive index is maximum. Equation (8.23) gives a central index of 1.406, which occurs at $z = z_0$, and an edge index of 1.386. Gullstrand specified the lens thickness as 3.6 mm. The refractive index profile contains purely axial and transverse changes in index and terms containing both y and z.

Finally, Pierscionek and Chan suggested that in the human lens, the refractive index distribution may be better represented by up to a sixth–order term. In such a distribution, the index value is relatively constant in the central region, and a gradient index exists only in the cortex [8.25].

8.4 The Gradient Parameter: Axial and Field Rays in the Crystalline Lens

In order to apply the results obtained for GRIN lenses in Chap. 3, we can now write the refractive index profile of the crystalline lens in the sagittal section, for the paraxial region as

$$n(y,z) = n_0(z)\left[1 - \frac{g^2(z)}{2}y^2\right] \tag{8.24}$$

where $n_0(z)$ is the index along the optical axis, and $g(z)$ is the gradient parameter that characterizes the refractive index distribution. All terms of higher order than the second in y are neglected. These higher order terms are aberration terms.

For the front part of the lens, we have that

$$g_f^2(z) = -\frac{\sum_{j=1}^{\infty} 2jc_j\left[\frac{z-a_1}{a_1}\right]^{2(j-1)}}{b^2 n_{0f}(z)} \tag{8.25}$$

where $n_{0f}(z)$ is given by Eq. (8.17a), and for the back part of the lens

$$g_b^2(z) = -\frac{\sum_{j=1}^{\infty} 2jc_j\left[\frac{z}{a_2}\right]^{2(j-1)}}{b^2 n_{0b}(z)} \tag{8.26}$$

where $n_{0b}(z)$ is given by Eq. (8.17b). The gradient parameter can also be expressed as

$$g_f^2(z) = -\frac{1}{a_1}\left(\frac{z-a_1}{b}\right)^2 \frac{d}{dz}[\ln n_{0f}(z)] \tag{8.27a}$$

$$g_b^2(z) = -\frac{1}{a_2}\left(\frac{z}{b}\right)^2 \frac{d}{dz}[\ln n_{0b}(z)] \tag{8.27b}$$

The values of $n_{0\binom{f}{b}}$ at the center of the optical axis and at the vertex of the front and back surfaces of the lens coincide with Eq. (8.11). The values of the gradient parameter at these positions are given by

$$g_c^2 = g_f^2(a_1) = g_b^2(0) = -\frac{2c_1}{b^2 c_0} = -\frac{2c_1}{n_c b^2} \tag{8.28a}$$

$$g_e^2 = g_f^2(0) = g_b^2(a_2) = -\frac{\sum_{j=1}^{\infty} 2jc_j}{b^2 n_e} \tag{8.28b}$$

where n_e is the edge index.

In visual optics, the refractive index profile in the crystalline lens of the eye is often regarded as parabolic, that is, all c_j coefficients in Eqs. (8.10) or (8.13) except c_0 and c_1 are zero. In this case, the general expressions for the gradient parameter at the front and back parts of the lens are given by

$$g_f^2(z) = -\frac{2c_1}{b^2\left[c_0 + c_1\left(\frac{z-a_1}{a_1}\right)^2\right]} \tag{8.29a}$$

$$g_b^2(z) = -\frac{2c_1}{b^2\left[c_0 + c_1\left(\frac{z}{a_2}\right)^2\right]} \tag{8.29b}$$

such that

$$g_e^2 = -\frac{2c_1}{b^2(c_0 + c_1)} = -\frac{2c_1}{n_e b^2} \tag{8.30}$$

where Eqs. (8.26), (8.28b), (8.17), and (8.11) have been used.

To estimate values of the gradient parameter, the Gullstrand model for the GRIN refractive index is used. The refractive index distribution has central and edge refractive indices of 1.406 and 1.386, respectively. The equatorial semidiameter of the lens is $b = 5\ \text{mm}$. Substitution of these values into Eqs. (8.28) provides the extreme values of the gradient parameter

$$g_c = 0.03373\ \text{mm}^{-1}; \quad g_e = 0.03397\ \text{mm}^{-1} \tag{8.31}$$

It is clear that the values of the gradient parameter depend on the coefficients c_j, which are the coefficients of the power series in the refractive index profile. The variation of the gradient parameter and the refractive index along the axis is shown in Fig. 8.5.

The axial and field rays in the crystalline lens can be found if condition (1.94) is fulfilled, that is

$$\frac{|\dot{g}(z)|}{g^2(z)} << 1 \tag{8.32}$$

where $\dot{g}$ denotes the derivative of g with respect to z.

In the lens we have that

$$\frac{|\dot{g}(z)|}{g^2(z)} = \begin{cases} \dfrac{2b}{a_1}\dfrac{\left|\sum_{j=0}^{\infty} c_j\left(\frac{z-a_1}{a_1}\right)^{2j}\sum_{j=2}^{\infty} j(j-1)c_j\left(\frac{z-a_1}{a_1}\right)^{2j-3} - \sum_{j=1}^{\infty} jc_j\left(\frac{z-a_1}{a_1}\right)^{2j-2}\sum_{j=1}^{\infty} jc_j\left(\frac{z-a_1}{a_1}\right)^{2j-1}\right|}{\left[\sum_{j=0}^{\infty} c_j\left(\frac{z-a_1}{a_1}\right)^{2j}\right]^{1/2}\left[-\sum_{j=1}^{\infty} 2jc_j\left(\frac{z-a_1}{a_1}\right)^{2j-2}\right]^{3/2}} & \text{for the front part} \\[2ex] \dfrac{2b}{a_2}\dfrac{\left|\sum_{j=0}^{\infty} c_j\left(\frac{z}{a_2}\right)^{2j}\sum_{j=2}^{\infty} j(j-1)c_j\left(\frac{z}{a_2}\right)^{2j-3} - \sum_{j=1}^{\infty} jc_j\left(\frac{z}{a_2}\right)^{2j-2}\sum_{j=1}^{\infty} jc_j\left(\frac{z}{a_2}\right)^{2j-1}\right|}{\left[\sum_{j=0}^{\infty} c_j\left(\frac{z}{a_2}\right)^{2j}\right]^{1/2}\left[-\sum_{j=1}^{\infty} 2jc_j\left(\frac{z}{a_2}\right)^{2j-2}\right]^{3/2}} & \text{for the back part} \end{cases} \tag{8.33}$$

Figure 8.6 depicts variation of Eq. (8.33) with z in the lens for two, three, and four coefficients of the power series in the refractive index distribution. The more coefficients we take into account, the better condition (8.32) is satisfied.

Therefore, the position and the slope of the axial and field rays in the crystalline lens can be expressed as

$$H_{af}(z) = [g_e g_f(z)]^{-1/2}\sin\left[\int_0^z g_f(z')dz'\right] = -[g_e g_f(z)]^{-1}\dot{H}_{ff}(z) \tag{8.34a}$$

$$H_{ff}(z) = \left[\frac{g_e}{g_f(z)}\right]^{1/2}\cos\left[\int_0^z g_f(z')dz'\right] = \frac{g_e}{g_f(z)}\dot{H}_{af}(z) \tag{8.34b}$$

for the front part, and

$$H_{ab}(z) = [g_e g_b(z)]^{-1/2}\sin\left[\int_0^{a_1+z} g_b(z')dz'\right] = -[g_e g_b(z)]^{-1}\dot{H}_{fb}(z) \tag{8.34c}$$

$$H_{fb}(z) = \left[\frac{g_e}{g_b(z)}\right]^{1/2}\cos\left[\int_0^{a_1+z} g_b(z')dz'\right] = \frac{g_e}{g_b(z)}\dot{H}_{ab}(z) \tag{8.34d}$$

for the back part such that

$$H_{af}(0) = 0; \quad H_{af}(a_1) = H_{ab}(0) \tag{8.35a}$$

$$H_{ff}(0) = 1; \quad H_{ff}(a_1) = H_{fb}(0) \tag{8.35b}$$

where Eqs. (1.92) and (1.95) have been used.

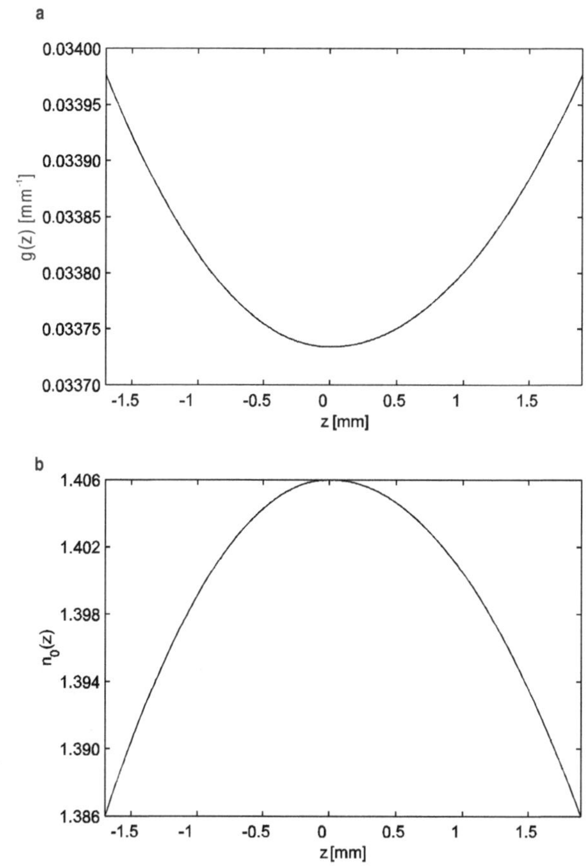

Fig. 8.5. Variation of the gradient parameter **a** and the refractive index **b** along the z axis. Calculations have been made for $c_0 = 1.406$, $c_1 = -0.02$, and $d = 3.6\,\text{mm}$ ($a_1 = 1.7\,\text{mm}$, $a_2 = 1.9\,\text{mm}$).

On the other hand, the gradient parameter changes very slowly with z, and we can expect that $g(z_1)z_1$ at z_1 is likely to be in $g(z_2)z_2$ at a very close z_2. Then, the integral in Eqs. (8.34) can be actually approximated by g(z)z.

Finally, g(z)z is a small quantity in the crystalline lens. Figure 8.7 shows the variation of g(z)z with z for the Gullstrand model. The maximum value of g(z)z is achieved at the vertex of the back surface of the lens and is about 0.12 rad $\approx 7^{\circ}$. For this value, approximations of first and second orders for the sine and cosine can be made, respectively.

Thus, the positions of the axial and field rays in the crystalline lens can be written as:

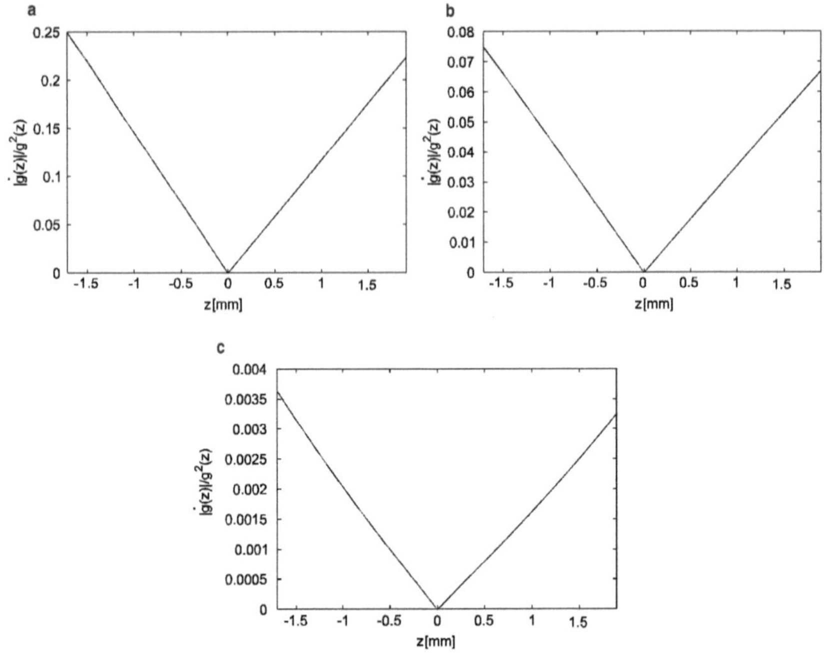

Fig. 8.6. Condition (8.32) versus z. Calculations have been made for $b = 5\,\text{mm}$, $a_1 = 1.7\,\text{mm}$, $a_2 = 1.9\text{mm}$, $c_0 = 1.406$ **a** $c_1 = -0.02$, **b** $c_1 = -0.0201$, $c_2 = 0.0001$; and **c** $c_1 = -0.0201416$, $c_2 = 0.0001423$, $c_3 = -0.0000007$.

$$H_{af}(z) = \left[\frac{g_f(z)}{g_e}\right]^{1/2} z; \quad H_{ff}(z) = \left[\frac{g_e}{g_f(z)}\right]^{1/2}\left[1 - \frac{g_f^2(z)}{2} z^2\right] \tag{8.36a}$$

for the front part, and

$$H_{ab}(z) = \left[\frac{g_b(z)}{g_e}\right]^{1/2} (z + a_1); \quad H_{fb}(z) = \left[\frac{g_e}{g_b(z)}\right]^{1/2}\left[1 - \frac{g_b^2(z)}{2}(z + a_1)^2\right] \tag{8.36b}$$

for the back part.

If Eqs. (8.34) are taken into account, the slope of the axial and field rays can be expressed as:

$$\dot{H}_{af}(z) = \left[\frac{g_f(z)}{g_e}\right]^{1/2}\left[1 - \frac{g_f^2(z)}{2} z^2\right]; \quad \dot{H}_{ff}(z) = -g_e^{1/2} g_f^{3/2}(z) z \tag{8.37a}$$

for the front part and

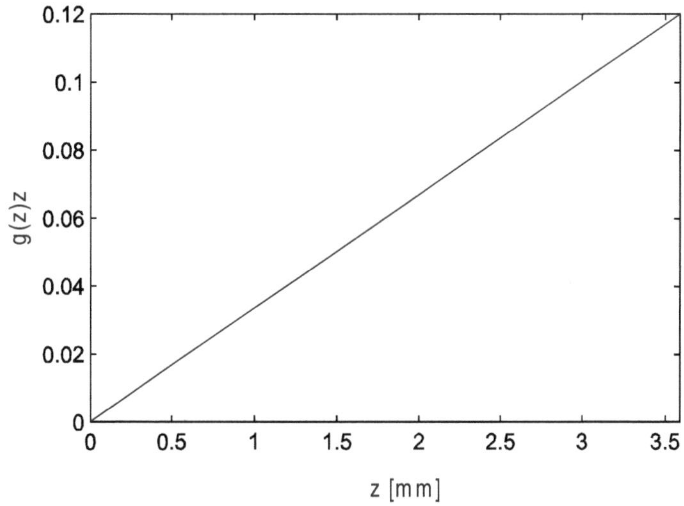

Fig. 8.7. Variation of g(z)z along the z–axis. Calculations have been made for b = 5mm, $c_0 = 1.406$, and $c_1 = -0.02$.

$$\dot{H}_{ab}(z) = \left[\frac{g_b(z)}{g_e}\right]^{1/2}\left[1 - \frac{g_b^2(z)}{2}(z + a_1)^2\right]; \dot{H}_{fb}(z) = -g_e^{1/2} g_b^{3/2}(z)(z + a_1) \quad (8.37b)$$

for the back part.

8.5 Refractive Power and Cardinal Points of the Crystalline Lens

The refractive power of the crystalline lens is dependent on the curvature radii of the front and back surfaces, the refractive index differences at the aqueous–lens and vitreous–lens interfaces, and the variation of the refractive index within the lens. The first two dependences give the refractive powers of the surfaces, which are evaluated from the standard surface power equation

$$P = \frac{n' - n}{r} \quad (8.38)$$

where r is the curvature radius of the surface, and n and n' are the refractive indices on the incident and refracted sides of the surface, respectively.

Here we are only concerned with the third dependence of the lens, which arises from its GRIN nature. By ignoring the surface refractions we can regard the crystalline lens as a GRIN lens slab with parallel faces and thickness $d = a_1 + a_2$ (Fig. 8.8).

The refractive power of the lens is expressed as

$$P_G = \frac{n_1'}{f'} = \frac{n_1'}{H'F'} \tag{8.39}$$

and the back vertex power $P_{V'}$ is written as

$$P_{V'} = \frac{n_1'}{l'} = \frac{n_1'}{V'F'} \tag{8.40}$$

where n_1' is the vitreous refractive index, and f' and l' are the back focal and working distances given by Eqs. (3.66b) and (3.60), respectively.

Substitution of Eqs. (8.36b) and (8.37b) into Eqs. (3.66b) and (3.60) provides

$$f' = \frac{n_1'}{n_e g_e^2 d} = H'F' \tag{8.41}$$

$$l' = \frac{n_1'}{n_e g_e^2 d}\left(1 - \frac{g_e^2}{2}d^2\right) = f'\left(1 - \frac{g_e^2}{2}d^2\right) = V'F' \tag{8.42}$$

Thus the refractive power and the back vertex power of the crystalline lens are given by

$$P_G = n_e g_e^2 d = -\frac{2d}{b^2}\sum_{j=1}^{\infty} jc_j \tag{8.43}$$

$$P_{V'} = \frac{n_e g_e^2 d}{1 - \frac{g_e^2}{2}d^2} = -\frac{2n_e d\sum_{j=1}^{\infty} jc_j}{n_e b^2 + d^2\sum_{j=1}^{\infty} jc_j} \tag{8.44}$$

where Eq. (8.28b) has been used.

Equations (8.43–8.44) show the refractive power depends on the edge index, the thickness, and the gradient parameter of the lens. If the refractive index is constant along the optical axis and $c_j = 0$ for $j > 1$, Eq. (8.43) reduces to the power of the Wood lens.

Likewise, the positions of the back cardinal points can be found by use of Eqs. (8.41–8.42) and (3.64). The position of the back focus with respect to vertex V' is given by Eq. (8.42), and the position of the back principal point with respect to V' is expressed as

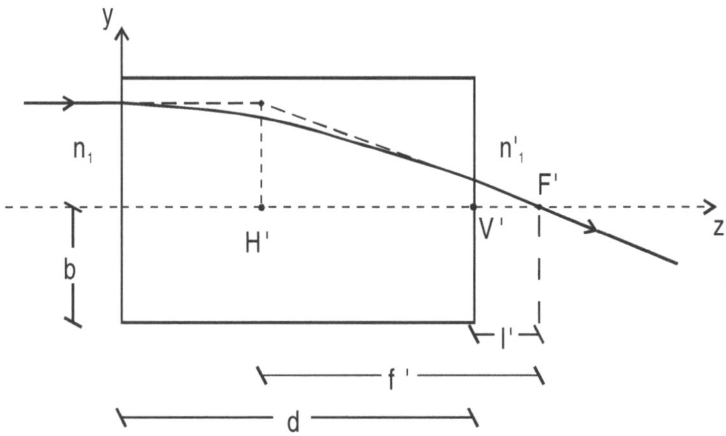

Fig. 8.8. The crystalline lens as a GRIN lens with parallel faces and thickness d.

$$V'H' = V'F' - H'F' = l' - f' = -\frac{n_1' d}{2n_e} \tag{8.45}$$

and there is no dependence on the gradient parameter.

From Eq. (3.64) the position of the back nodal point can be evaluated, that is,

$$V'N' = \frac{n_1 - n_1' H_{fb}(d)}{n_e \dot{H}_{fb}(d)} \tag{8.46}$$

where n_1 is the aqueous refractive index.

Inserting Eqs. (8.36b) and (8.37b) into Eq. (8.46) we have

$$V'N' = -\frac{n_1 - n_1'\left(1 - \frac{g_e^2}{2} d^2\right)}{n_e g_e^2 d} \tag{8.47}$$

Therefore Eq. (8.47) reduces to Eq. (8.45), and the nodal and principal points coincide as $n_1 = n_1'$. Aqueous and vitreous refractive indices have approximately the same value of 1.336.

In the same way, the positions of the front cardinal points can be calculated by use of Eqs. (3.59), (3.63), and (3.66a), which are written for the crystalline lens as

$$f = HF = -\frac{n_1}{n_e \dot{H}_{fb}(d)} = \frac{n_1}{n_e g_e^2 d} \tag{8.48}$$

$$1 = \mathrm{VF} = -\frac{n_1 \dot{H}_{ab}(d)}{n_e \dot{H}_{fb}(d)} = \frac{n_1}{n_e g_e^2 d}\left(1 - \frac{g_e^2 d^2}{2}\right) = f\left(1 - \frac{g_e^2 d^2}{2}\right) \tag{8.49}$$

$$\mathrm{VN} = \frac{n_1' - n_1 \dot{H}_{ab}(d)}{n_e \dot{H}_{fb}(d)} = -\frac{n_1' - n_1\left(1 - \frac{g_e^2 d^2}{2}\right)}{n_e g_e^2 d} \tag{8.50}$$

Equations (8.48–49) represent the front focal and working distances, respectively. They are equal to the corresponding back distances as $n_1 = n_1'$. Equation (8.50) gives the position of the front nodal point with respect to vertex V.

Furthermore, the position of the front principal point is given by

$$\mathrm{VH} = \mathrm{VF} - \mathrm{HF} = -\frac{n_1 d}{2 n_e} \tag{8.51}$$

From Eqs. (8.50–8.51) it follows that the front nodal and principal points coincide. It also follows that the front and back principal points are located at the same distance from the vertices V and V', respectively, as $n_1 = n_1'$.

Finally, if we know the power of the crystalline lens due to its GRIN nature and the position of its principal points, the refractive power of the whole lens is evaluated from the following equation [8.26]

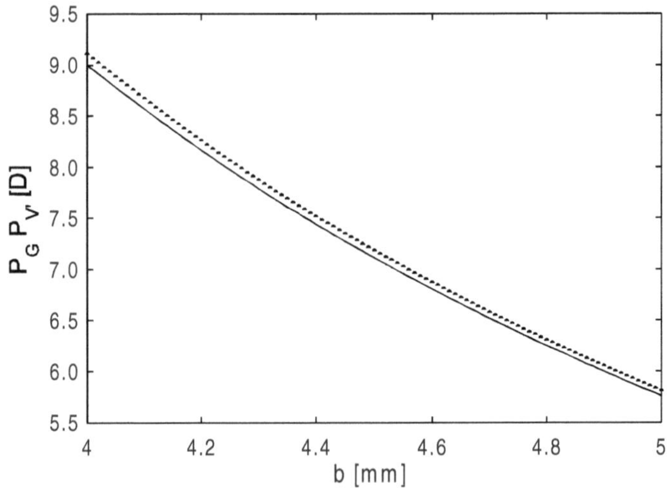

Fig. 8.9. The refractive power (solid line) and the back vertex power (dashed line) of the crystalline lens versus the equatorial radius. Calculations have been made for $n_e = 1.386$, $c_0 = 1.406$, $c_1 = -0.02$, and d = 3.6mm.

$$P_W = P_f + P_b + P_G - HV\frac{P_f P_G}{n_1} - H'V'\frac{P_b P_G}{n_1'} - \left(\frac{HV}{n_1} + \frac{H'V'}{n_1'}\right)P_f P_b + \frac{HV}{n_1}\cdot\frac{H'V'}{n_1'}P_f P_G P_b \quad (8.52)$$

where P_f and P_b are the powers of the front and back surfaces, respectively.

Let us now estimate the GRIN power, the position of the principal points, and the whole power of the lens of the eye in the Gullstrand model, provided that $n_1 = n_1' = 1.336$, $d = 3.6\ \text{mm}$, and $b = 5$ mm.

The GRIN power and the position of the principal points are given by

$$P_G = 5.76\text{D}; \quad P_{V'} = 5.80\text{D}; \quad H'V' = HV = 1.74\text{mm} \quad (8.53)$$

where Eqs. (8.43–8.45) have been used.

Figure 8.9 represents the variation of the refractive power and the back vertex power of the crystalline lens versus the equatorial radius. A strong dependence of the GRIN power on the equatorial radius is shown. The refractive power achieves a value of 6 D at $b = 4.9\ \text{mm}$.

On the other hand, front and back surface radii of curvature of the lens are $10\ \text{mm}$ $(\rho < 0)$ and –6mm $(\rho = 0)$, corresponding to a hyperboloid and paraboloid, respectively. These radii of curvature provide front surface power of $5\ \text{D}$ and back surface power of $-8.33\ \text{D}$. We can now use the above values to calculate the whole power of the lens, which is given by

$$P_W = 19.27\ \text{D} \quad (8.54)$$

From Eq. (8.54) it follows that the value of the whole power of the lens is close to 19 diopters for an unaccommodated lens [8.26–8.27].

9 Optical Connections by GRIN Lenses

9.1 Introduction

GRIN optics is an active area of research in optical fiber transmission systems and optical sensing. In particular, GRIN lenses provide on–axis and off–axis imaging and Fourier transforming, which are of great importance in these fields. Taking advantage of the inherent features of the lenses, many optical configurations for connecting fibers, sources, or detectors by GRIN lenses have been analyzed for coupling and sensing purposes [9.1–9.4]. We will study three typical configurations, that is, single–mode fibers coupled by GRIN fiber lenses, corrective elements of astigmatic Gaussian beams from laser diodes by anamorphic selfoc lenses, and beam–size controllers and light deflectors made from tapered GRIN lenses. Other devices using selfoc lenses are also presented.

This chapter shows some devices that can be made with GRIN lenses and gives ways in which they can be used in optical systems for processing and manipulating signals.

9.2 GRIN Fiber Lens

GRIN fiber lenses are used in monomode fiber communication systems for obtaining low–loss connections between monomode fibers with different spot sizes [9.5–9.6]. The GRIN fiber lenses are manufactured from multimode GRIN fibers with small diameters of about 100 μm, and gradient parameter values up to 6 mm^{-1} to achieve strong focusing. An optical connector consisting of a GRIN fiber lens fused to monomode fibers and aligned by surface tension is shown in Fig. 9.1.

Let us consider the mode profile of the monomode fibers as Gaussian distributions, and the refractive index profile of the GRIN fiber lens as Eq. (1.69). The Gaussian approximation to the fundamental mode–field profile of step–index single–mode fibers has been studied in detail. It has been shown that for the normalized parameter V larger than 2 (the cutoff of the next higher order mode occurs at V = 2.405), the Gaussian approximation is very accurate [9.7–9.8].

Our aim is to obtain a maximum coupling efficiency between the fundamental modes of both fibers by a scaling operation carried out by the GRIN fiber lens. The integral transformation performed by the GRIN fiber lens of thickness d is written as Eq. (2.1), that is

$$\psi(x,y;d)=\int_{\Re^2} K(x_0,y_0,x,y;d)\psi_1(x_0,y_0)\,dx_0dy_0 \tag{9.1}$$

where K is the kernel of the linear transformation given by Eqs. (2.20) or (2.21), and ψ_1 is the mode field of the first fiber.

An exact reconstruction of ψ_1 with a scaling factor on the input of the second monomode fiber is obtained as the image condition is fulfilled. From Eq. (2.25) it follows that we can obtain the exact reconstruction of the mode field ψ_1 with scaling factor

$$A=(-1)^m\left[\frac{g_0}{g(z_m)}\right]^{1/2} \tag{9.2}$$

by a GRIN fiber lens of thickness z_m, if the following condition is satisfied

$$H_a(z_m)=\dot{H}_f(z_m)=0 \tag{9.3}$$

where m is an integer.

Therefore, the GRIN fiber lens connector provides an appropriate scaling operation of mode fields for optimizing the mode coupling. Now we present a simple and useful case of coupling by a GRIN fiber lens. Let us consider two elliptical Gaussian mode fields with slightly different eccentricities for coupling between single–mode fibers that preserve polarization [9.9–9.11]. The normalized mode fields are given by

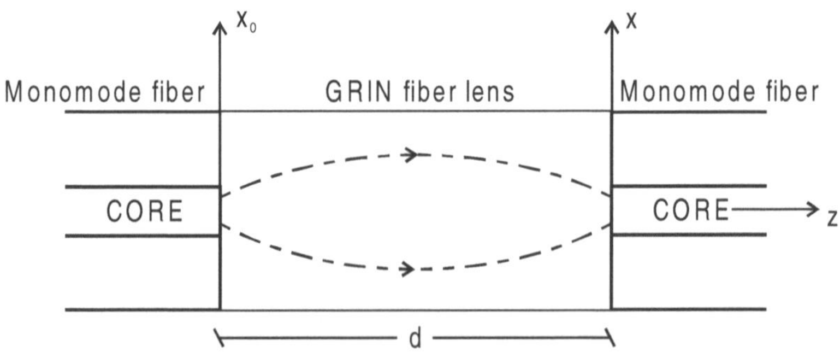

Fig. 9.1. Optical connector implemented by a GRIN fiber lens of thickness d fused to the end of monomode fibers.

$$\psi_1(x_0, y_0) = \sqrt{\frac{2}{\pi w_{x1} w_{y1}}} \exp\left\{-\left(\frac{x_0^2}{w_{x1}^2} + \frac{y_0^2}{w_{y1}^2}\right)\right\} \tag{9.4a}$$

$$\psi_2(x, y) = \sqrt{\frac{2}{\pi w_{x2} w_{y2}}} \exp\left\{-\left(\frac{x^2}{w_{x2}^2} + \frac{y^2}{w_{y2}^2}\right)\right\} \tag{9.4b}$$

where w_{xi} and w_{yi} ($i = 1,2$) are the halfwidths of the mode fields along the x– and y–axes respectively.

After the scaling operation by the GRIN fiber lens, the first mode field on the input of the second fiber can be expressed as

$$\psi_1(x, y) = \frac{1}{A}\sqrt{\frac{2}{\pi w_{x1} w_{y1}}} \exp\left\{-\frac{1}{A^2}\left(\frac{x^2}{w_{x1}^2} + \frac{y^2}{w_{y1}^2}\right)\right\} \tag{9.5}$$

where Eq. (2.28) has been used, and a constant phase factor has been omitted.

The coupling efficiency between the field distributions is given by the overlap integral

$$\eta = \left| \int_{\Re^2} \psi_1(x, y)\psi_2^*(x, y)dxdy \right|^2 \tag{9.6}$$

A straightforward calculation leads to the following expression for η

$$\eta = \frac{4A^2 w_{x1} w_{x2} w_{y1} w_{y2}}{(w_{x2}^2 + A^2 w_{x1}^2)(w_{y2}^2 + A^2 w_{y1}^2)} \tag{9.7}$$

Therefore, the maximum value of η for a given A can be evaluated by the extreme condition

$$\frac{d\eta}{dA} = 0 \tag{9.8}$$

Solving the extreme condition and taking into account Eq. (9.7), the value of the scaling factor is given by

$$A^2 = \frac{w_{x2} w_{y2}}{w_{x1} w_{y1}} \tag{9.9}$$

at which the maximum coupling efficiency is expressed as

$$\eta_{max} = \frac{4 w_{x1} w_{x2} w_{y1} w_{y2}}{(w_{x1} w_{y2} + w_{y1} w_{x2})^2} \tag{9.10}$$

Thus for the GRIN fiber lens connector, the maximum efficiency is given by

$$\eta_{max} = \frac{4g_0}{g(z_m)\left[\frac{w_{x2}}{w_{x1}} + \frac{w_{y2}}{w_{y1}}\right]^2} \tag{9.11}$$

as the condition

$$\frac{w_{x2} w_{y2}}{w_{x1} w_{y1}} = \frac{g_0}{g(z_m)} \tag{9.12}$$

is fulfilled, where Eq. (9.2) has been used.

From Eq. (9.12) it follows that, for instance, two identical monomode fibers can be connected by a half–pitch selfoc fiber lens of thickness $d = \pi / g_0$ with maximum efficiency as

$$w_{x1} w_{y1} = w_{x2} w_{y2} \tag{9.13}$$

A nonscaled image of the input mode field is obtained. This is a well–known result in the literature about axial coupling between elliptical Gaussian modes.

9.3 Anamorphic Selfoc Lens

Laser diodes offer significant advantages over other laser sources in efficiency, size, and cost, but suffer from inferior optical characteristics. Their beams diverge, have asymmetric cross sections, and are often astigmatic. These deficiencies must be corrected to comply with many of the current applications of laser diodes in communications, data storage, and imaging. Integrated–optics lenses or gratings and holographic elements have been proposed as corrective elements [9.12–9.13].

Here we are concerned with anamorphic selfoc lenses to correct astigmatic Gaussian beams from laser diodes (Fig. 9.2). On the one hand, we consider an anamorphic selfoc lens whose refractive index profile is given by

$$n^2(x, y) = n_0^2\left(1 - g_x^2 x^2 - g_y^2 y^2\right) \tag{9.14}$$

where n_0 is the index along the optical axes z, and g_x and g_y are the gradient parameters along the x– and y–axes describing the evolution of the transverse parabolic index distribution. The equi–index surfaces are cylinders of elliptical basis and axis z.

On the other hand, at a distance d_1 from the output face of the laser diode, the astigmatic beam can be expressed as a Gaussian beam [9.14]

$$\psi(x_0, y_0) = \sqrt{\frac{w_{0x} w_{0y}}{w_x(d_1) w_y(d_1)}} \exp\{i\varphi(d_1)\} \exp\left\{i\frac{\pi}{\lambda}\left[U_x(d_1)x_0^2 + U_y(d_1)y_0^2\right]\right\} \tag{9.15}$$

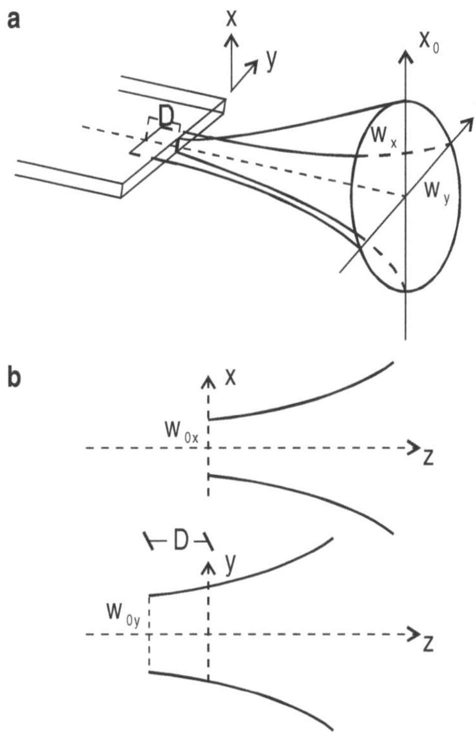

Fig. 9.2. Astigmatic Gaussian beam produced by a laser diode **a** D is the axial astigmatism, and **b** the difference between the waist w_{0x} and w_{0y} is referred to as transversal astigmatism.

where the beam parameters are given by the complex wavefront curvatures

$$U_{\binom{x}{y}}(d_1) = \frac{1}{R_{\binom{x}{y}}(d_1)} + i\frac{\lambda}{\pi w^2_{\binom{x}{y}}(d_1)} \tag{9.16}$$

and the on–axis phase

$$\varphi(d_1) = \frac{1}{2}\left[\tan^{-1}\left(\frac{\lambda(d_1 + D)}{\pi w_{0x}^2}\right) + \tan^{-1}\left(\frac{\lambda d_1}{\pi w_{0y}^2}\right)\right] \tag{9.17}$$

where $R_{\binom{x}{y}}$ and $w_{\binom{x}{y}}$ are the principal radii of curvature and the beam half–widths at d_1, respectively, and D is the longitudinal astigmatism of the laser beam. The emitted beam has its waist w_{0x} and w_{0y}, both perpendicular and parallel to the junction, at the output face of the laser diode, as shown in Fig. 9.2.

Likewise, the relationship between the waists and the half–widths at d_1 and the radii of curvature is given by

$$w_y^2(d_1) = w_{0y}^2\left[1+\left(\frac{(d_1+D)\lambda}{\pi w_{0y}^2}\right)^2\right]; \quad w_x^2(d_1) = w_{0x}^2\left[1+\left(\frac{d_1\lambda}{\pi w_{0x}^2}\right)^2\right] \tag{9.18a}$$

$$R_y(d_1) = (d_1+D)\left[1+\left(\frac{\pi w_{0y}^2}{(d_1+D)\lambda}\right)^2\right]; \quad R_x(d_1) = d_1\left[1+\left(\frac{\pi w_{0x}^2}{d_1\lambda}\right)^2\right] \tag{9.18b}$$

Note that when $D = 0$ and $w_{0x} = w_{0y}$, Eq. (9.15) becomes the complex amplitude distribution of a spherical Gaussian beam given by Eq. (4.1).

If Eq. (9.15) denotes the mode field on the input face of the anamorphic selfoc lens of thickness d, the field at the output face can be easily evaluated by Eq. (2.1), where the kernel of the integral equation is now separable in both transverse axes and is given by

$$K(x_0, y_0, x', y'; d) = \frac{kn_0 \exp\{ikn_0 d\}}{i2\pi[H_{ax}(d)H_{ay}(d)]^{1/2}} \exp\left\{i\frac{kn_0}{2H_{ax}(d)}[x'^2\dot{H}_{ax}(d) + x_0^2 H_{fx}(d) - 2x'x_0]\right\}$$
$$\exp\left\{i\frac{kn_0}{2H_{ay}(d)}[y'^2\dot{H}_{ay}(d) + y_0^2 H_{fy}(d) - 2y'y_0]\right\} \tag{9.19}$$

where $H_{a\binom{x}{y}}$ and $H_{f\binom{x}{y}}$ are the axial and field rays in both transverse directions, that is,

$$H_{a\binom{x}{y}}(d) = \frac{\sin\left[g_{\binom{x}{y}} d\right]}{g_{\binom{x}{y}}} = -\frac{\dot{H}_{f\binom{x}{y}}(d)}{g^2_{\binom{x}{y}}} \tag{9.20a}$$

$$H_{f\binom{x}{y}}(d) = \cos\left[g_{\binom{x}{y}} d\right] = \dot{H}_{a\binom{x}{y}}(d) \tag{9.20b}$$

Substituting Eqs. (9.15–9.16) and (9.19) into Eq. (2.1) and performing the integration, the complex amplitude distribution at the output face of the anamorphic selfoc lens can be expressed as [9.15]

$$\psi(x', y'; d) = \sqrt{\frac{w_{0x} w_{0y}}{\Omega_x(d)\Omega_y(d)}} \exp\{i\varphi(d)\} \exp\left\{i\frac{\pi}{\lambda}[U_x(d)x'^2 + U_y(d)y'^2]\right\} \tag{9.21}$$

where the complex curvatures and the on–axis phase are given by

$$U_{\binom{x}{y}}(d) = \frac{1}{\rho_{\binom{x}{y}}(d)} + i\frac{\lambda}{\pi\Omega^2_{\binom{x}{y}}(d)} \tag{9.22}$$

$$\begin{aligned}\varphi(d) = kn_0 d + \varphi(d_1) + \frac{1}{2}\Bigg[\tan^{-1}\left(\frac{\pi w_x^2(d_1)[H_{ax}(d) + n_0 R_x(d_1) H_{fx}(d)]}{\lambda R_x(d_1) H_{ax}(d)}\right) \\ + \tan^{-1}\left(\frac{\pi w_y^2(d_1)[H_{ay}(d) + n_0 R_y(d_1) H_{fy}(d)]}{\lambda R_y(d_1) H_{ay}(d)}\right)\Bigg]\end{aligned} \tag{9.23}$$

and the radii of curvature and the beam half–widths are written as

$$\begin{aligned}\rho_{\binom{x}{y}}(d) &= \frac{\left[H_{a\binom{x}{y}}(d) + n_0 H_{f\binom{x}{y}}(d) R_{\binom{x}{y}}(d_1)\right]^2 + \left[\dfrac{\lambda R_{\binom{x}{y}}(d_1) H_{a\binom{x}{y}}(d)}{\pi w^2_{\binom{x}{y}}(d_1)}\right]^2}{\dot{H}_{a\binom{x}{y}}(d) H_{a\binom{x}{y}}(d)\left[\dfrac{\lambda R_{\binom{x}{y}}(d_1)}{\pi w^2_{\binom{x}{y}}(d_1)}\right]^2 + \left[n_0 R_{\binom{x}{y}}(d_1)\dot{H}_{f\binom{x}{y}}(d) + \dot{H}_{a\binom{x}{y}}(d)\right]\left[n_0 R_{\binom{x}{y}}(d_1) H_{f\binom{x}{y}}(d) + H_{a\binom{x}{y}}(d)\right]} \\ &= \left\{\frac{1}{2}\frac{d}{dz}\ln\left[\left(H_{a\binom{x}{y}}(z) + n_0 H_{f\binom{x}{y}}(z) R_{\binom{x}{y}}(d_1)\right)^2 + \left(\frac{\lambda R_{\binom{x}{y}}(d_1) H_{a\binom{x}{y}}(z)}{\pi w^2_{\binom{x}{y}}(d_1)}\right)^2\right]\Bigg|_{z=d}\right\}^{-1}\end{aligned} \tag{9.24}$$

$$\Omega^2_{\binom{x}{y}}(d) = w^2_{\binom{x}{y}}(d_1)\left[\left(H_{f\binom{x}{y}}(d) + \frac{H_{a\binom{x}{y}}(d)}{n_0 R_{\binom{x}{y}}(d_1)}\right)^2 + \left(\frac{\lambda H_{a\binom{x}{y}}(d)}{\pi n_0 w^2_{\binom{x}{y}}(d_1)}\right)^2\right] \tag{9.25}$$

So, in order to correct the astigmatism in a laser diode by this kind of lens, the index along the axis n_0, gradient parameters g_x and g_y, and thickness d of the lens must be chosen in such a way that

$$\rho_x = \rho_y \tag{9.26}$$

to correct longitudinal astigmatism, and

$$\Omega_x = \Omega_y \tag{9.27}$$

to correct transverse astigmatism at any plane behind the anamorphic selfoc lens. By solving the above set of equations, one can obtain a rotationally symmetric

Gaussian beam at waist w_0 located at distance d_1' from the output face of the GRIN lens, as shown in Fig. 9.3.

Note that these results contain the conditions for correcting astigmatism for uniform illumination by performing the limits: $\lim_{w_x \to \infty} \rho_x$ and $\lim_{w_x \to \infty} \rho_y$. When $w_{\binom{x}{y}} \to \infty$, $\Omega_{\binom{x}{y}} \to \infty$ and the Gaussian amplitude of the emerging beam from the lens tends toward unity [9.16].

Some applications require only circular symmetry of the beam, without regard to position and size of the output beam waist. For this applications it is interesting to study a particular design solution for the anamorphic selfoc lens in which the set of transcendentals given by Eqs. (9.26–9.27) presents an analytical solution that takes into account the following conditions: (i) the lens must be attached to the output face of the laser diode so that one of its transverse directions fits with the direction perpendicular to the junction where the waist w_{0x} of the emitted beam is located, and (ii) the index along the axis and the gradient parameter in the x–direction must be verified

$$w_{0x} = \left[\frac{\lambda}{\pi n_0 g_x}\right]^{1/2} \tag{9.28}$$

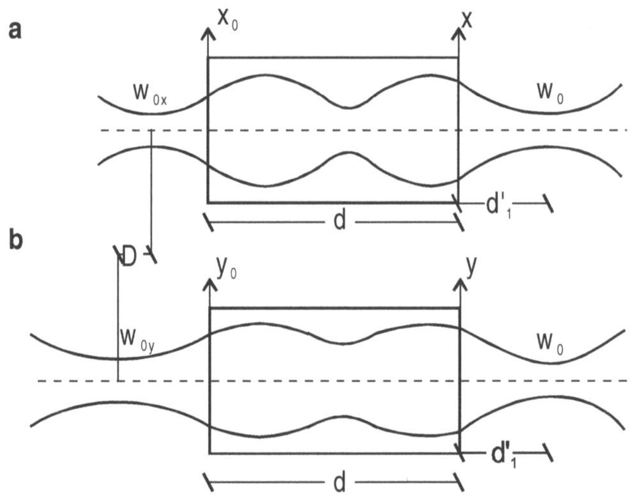

Fig. 9.3. Anamorphic selfoc lens to correct the astigmatism in a diode laser: **a** direction perpendicular to the junction, **b** direction parallel to the junction.

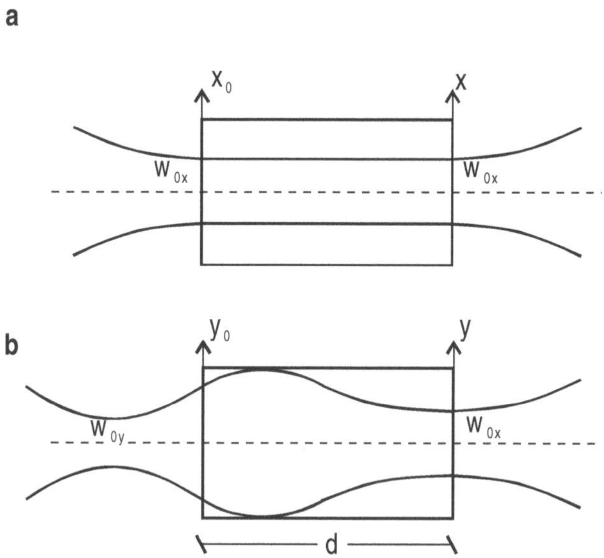

Fig. 9.4. Anamorphic selfoc lens to correct the astigmatism in a laser diode by matching to the fundamental mode: **a** direction perpendicular to the junction, **b** direction parallel to the junction.

With this in mind, the waist emitted in the x–direction matches the fundamental mode of the lens and propagates with distortion as discussed in Sect. 4.2. Thus

$$\rho_x \to \infty \quad ; \quad \Omega_x = w_{0x} \tag{9.29}$$

for any value of the thickness of the lens.

Since a spherical Gaussian beam of waist w_{0x} must be obtained at the output face of the anamorphic lens, the thickness d and the gradient parameter g_y can be evaluated by the following conditions

$$\rho_y \to \infty \quad \text{or} \quad \left.\frac{d\Omega_y^2}{dz}\right|_{z=d} = 0 \tag{9.30}$$

$$\Omega_y = w_{0x} \tag{9.31}$$

From Eq. (9.30) it follows that

$$\left[H_{fy}(d) + \frac{H_{ay}(d)}{n_0 R_y(d_1)}\right]\left[\dot{H}_{fy}(d) + \frac{\dot{H}_{ay}(d)}{n_0 R_y(d_1)}\right] = -\frac{\lambda^2 H_{ay}(d)\dot{H}_{ay}(d)}{\pi^2 n_0^2 w_y^4(d_1)} \tag{9.32}$$

where Eq. (9.25) has been used.

Under the change of variable

$$u_y = \tan[g_y d] \tag{9.33}$$

that is,

$$H_{ay}(d) = -\frac{\dot{H}_{fy}(d)}{g_y^2} = \frac{u_y}{g_y[1+u_y^2]^{1/2}} \tag{9.34a}$$

$$H_{fy}(d) = \dot{H}_{ay}(d) = \frac{1}{[1+u_y^2]^{1/2}} \tag{9.34b}$$

Solution of Eq. (9.32), taking into account Eq. (9.28), is given by

$$u_y = \frac{n_0 R_y(d_1)}{2g_y}\left[\left(\frac{w_{0x}}{w_y(d_1)}\right)^4 g_x^2 + \frac{1}{n_0^2 R_y^2(d_1)} - g_y^2\right] - \left[\frac{n_0^2 R_y^2(d_1)}{4g_y^2}\left\{\left(\frac{w_{0x}}{w_y(d_1)}\right)^4 g_x^2 + \frac{1}{n_0^2 R_y^2(d_1)} - g_y^2\right\}^2 + 1\right]^{1/2} \tag{9.35}$$

Note that the negative sign for the square root in Eq. (9.35) has been taken for obtaining the minimum value of the beam half–width that occurs when $w_{0x} > w_{0y}$ or $w_{0x} < w_{0y}$ and the lens is illuminated by a divergent beam. Equation (9.35) indicates that a value of u_y can be obtained for a given g_y, and from Eq. (9.33) the thickness d of the lens will be evaluated. In this way, we have a waist of size w_{0x} at the output face of the anamorphic selfoc lens, as shown in Fig. 9.4.

9.4 Tapered GRIN Lens

In applications requiring beam–size control and light deflection for coupling purposes, the use of tapered GRIN lenses offers advantages over conventional lenses in efficiency, small size, and low cost. A simple study of light propagation by geometrical optics in a tapered GRIN lens illuminated by a tilted plane beam will be presented. This section shows the performance of an optical component for controlling beam size and for deflecting incident light.

We will consider a tapered GRIN lens of thickness d and semiaperture a whose refractive index profile is given by Eq. (3.1). When the lens is illuminated by a tilted plane beam with input slope $\dot{r}_0$, not all rays reaching the input face will be confined through it. It is necessary to find the input effective semiaperture for the upper and lower marginal rays, as discussed in Sect. 3.4. Figure 9.5 shows the trajectory of the upper and lower marginal rays of the tilted plane beam that impinges on the lens. At a distance d_u from the input face of the lens where the upper marginal ray suffers the maximum deviation with respect to the axis, the position and the slope of this ray can be written as

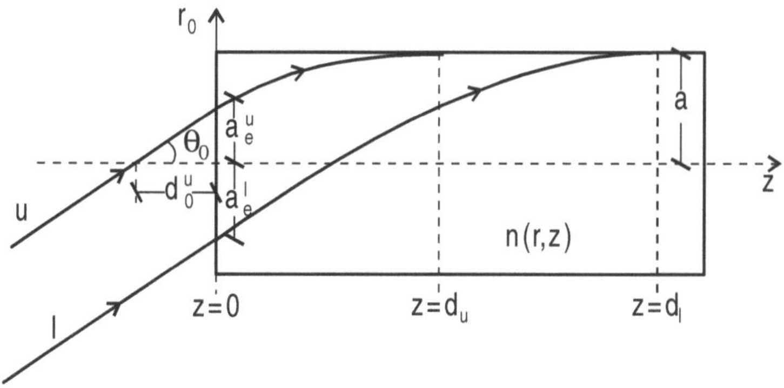

Fig. 9.5. Effective semiaperture for the upper and the lower marginal rays in a tapered GRIN lens illuminated by a tilted plane beam.

$$r(d_u) = a_e^u F(d_u) = a \tag{9.36}$$

$$\dot{r}(d_u) = a_e^u \dot{F}(d_u) = 0 \Rightarrow \dot{F}(d_u) = 0 \tag{9.37}$$

where

$$\overset{(\cdot)}{F}(d_u) = \overset{(\cdot)}{H}_f(d_u) + \frac{\overset{(\cdot)}{H}_a(d_u)}{n_0 d_0^u} \tag{9.38}$$

where d_0^u is the distance from the input face of the lens to the cutoff point of the upper marginal ray with the axis, and a_e^u is the upper effective semiaperture (Fig.9.5).

Then, the upper effective semiaperture is given by

$$a_e^u = a\dot{H}_a(d_u) \tag{9.39}$$

where Eqs. (9.36–9.37) and (1.79b) have been used.

Likewise, the input slope can be expressed as

$$\dot{r}_0 = \frac{\theta_0}{n_0} = \frac{a_e^u}{n_0 d_0^u} = -a\dot{H}_f(d_u) \tag{9.40}$$

where Eqs. (9.37–9.39) have been considered.

From Eqs. (9.39–9.40) it follows that the effective semiaperture and the input slope for the upper marginal ray are proportional to the slope of the axial and field rays at d_u, respectively.

Equation (9.39) can be written, in terms of the taper function g(z) and the input slope, as

$$a_e^u = a\left[\frac{g(d_u)}{g_0} - \left(\frac{\dot{r}_0}{ag_0}\right)^2\right]^{1/2} \tag{9.41}$$

where trigonometric relationships and Eqs. (9.37), (1.92), and (1.95) have been used, and g_0 is the value of g(z) at the input face of the lens.

In the same way, we can evaluate the effective semiaperture for the lower marginal ray (Fig. 9.5). In this case, we have

$$a_e^l = -a\dot{H}_a(d_l) = -a\left[\frac{g(d_l)}{g_0} - \left(\frac{\dot{r}_0}{ag_0}\right)^2\right]^{1/2} \tag{9.42}$$

where

$$\dot{r}_0 = -a\dot{H}_f(d_l) \tag{9.43}$$

and d_l is the tapered GRIN lens length for which the slope of the lower marginal ray is parallel to the z–axis.

On the other hand, for the imaging condition given by Eq. (9.3), the upper and lower marginal ray equations at z_m lens lengths become

$$r^{\binom{u}{l}}(z_m) = a_e^{\binom{u}{l}} H_f(z_m) \tag{9.44}$$

$$\dot{r}^{\binom{u}{l}}(z_m) = \dot{r}_0 \dot{H}_a(z_m) \tag{9.45}$$

Equations (9.44–9.45) show that the position and the slope of the marginal rays at z_m are proportional, respectively, to the position and the slope of the input. In other words, input rays, all of one direction, produce a parallel output beam of rays in some other direction whose scaling factor for position (transverse magnification) and for slope (angular magnification) are given by

$$m_t = H_f(z_m) = (-1)^m\left[\frac{g_0}{g(z_m)}\right]^{1/2} \tag{9.46}$$

$$m_a = \dot{H}_a(z_m) = (-1)^m\left[\frac{g(z_m)}{g_0}\right]^{1/2} \tag{9.47}$$

where Eqs. (9.44–9.45) have been used.

Thus, cutting the tapered GRIN lens at z_m, lengths we can obtain an optical component for changing the light propagation direction and for controlling the size of a light beam [9.17].

We will apply these results to a tapered GRIN lens whose taper function is given by

$$g(z) = \frac{g_0}{1 - \left(\frac{z}{L}\right)^2} \tag{9.48}$$

In this case, the equi–index surfaces for the refractive index are revolution paraboloids around the z–axis with common apices at $z = \pm L$. Axial and field rays can be written as [9.11, 9.18]

$$H_a(z) = \frac{(L^2 - z^2)^{1/2}}{g_0 L} \sin\left[g_0 L \tanh^{-1}\left(\frac{z}{L}\right)\right] \tag{9.49a}$$

$$H_f(z) = \left(1 - \frac{z^2}{L^2}\right)^{1/2} \cos\left[g_0 L \tanh^{-1}\left(\frac{z}{L}\right)\right] \tag{9.49b}$$

and the z_m lengths verifying the imaging condition are given by

$$z_m = L \tanh\left(\frac{m\pi}{g_0 L}\right) \tag{9.50}$$

For this profile, the input effective semiaperture for the upper and lower marginal rays and the output beam slope can be expressed as

$$a_e^{\binom{u}{l}} = \pm a \left[\frac{L^2}{L^2 - d^2_{\binom{u}{l}}} - \left(\frac{\dot{r}_0}{a g_0}\right)^2\right]^{1/2} \tag{9.51}$$

$$\dot{r}_e = \frac{\theta_e}{n_0} = (-1)^m \dot{r}_0 \cosh\left(\frac{m\pi}{g_0 L}\right) \tag{9.52}$$

where θ_e is the deflection angle of the output beam and where Eqs. (9.41–9.42), (9.45), (9.47–9.48) and (9.50) have been used.

Figure 9.6a shows the dependence of the effective semiaperture on the input beam slope for this kind of lens. As expected, semiaperture decreases with the input slope. Figure 9.6b represents the hyperbolic variation of the deflection angle of the output beam with integer m for the first six zeros of H_a.

Figure 9.7 depicts a device designed by cutting a tapered GRIN lens at length z_1 for which the first zero of H_a is obtained. The deflection angle and the output beam size depend on integer m and on lens parameters. Output beam sizes of 678μm at z_1 and of 344.5 μm at z_2 with deflection angles of –6.8° and 13.3°, respectively, can be obtained when a tapered GRIN lens of parameters $a = 500\,\mu\text{m}$, $L = 20\,\text{mm}$, $g_0 = 0.192\,\text{mm}^{-1}$, and $n_0 = 1.6$ is illuminated by a tilted plane beam of $\theta_0 = 5°$.

Note that for the principle of reversibility of rays, this device can also be used as a beam expander and for beam–slope reduction.

Finally, when a collimated beam impinges normally on the input face of the tapered GRIN lens, the physical semiaperture a becomes the input effective semiaperture a_e. Therefore the emerging collimated beam propagates along the same direction as the input. The lens works as a beam–size controller to contract a collimated beam. The relationship between sizes of input and output beams is given by Eq. (9.46), and the lens can be applied for butt–joining coupling between two multimode fibers of different core sizes.

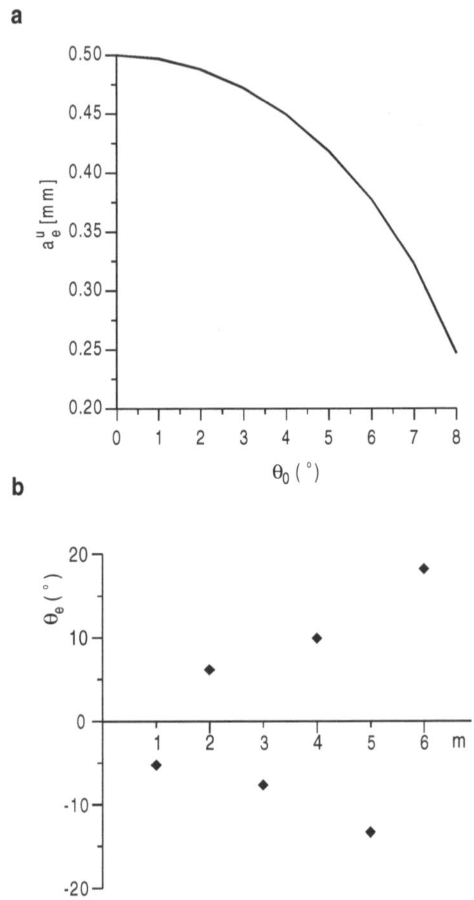

Fig. 9.6. Dependence of **a** upper effective semiaperture on the input beam slope, and **b** the deflection angle of the output beam on the number of the output plane m. Calculations have been made for $a = 500\,\mu m$, $g_0 = 0.192\,mm^{-1}$, $n_0 = 1.6$ and $L = 20mm$ (a), and $L = 50\,mm$ (b).

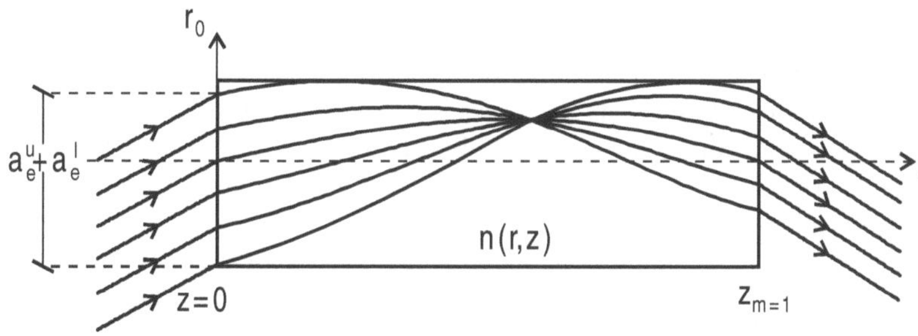

Fig. 9.7. Light deflector and beam size controller obtained cutting the lens at z_1. Ray–tracing has been made for $a = 500\,\mu m$, $L = 20\,mm$, $g_0 = 0.192\,mm^{-1}$, $n_0 = 1.6$, $\theta_0 = 5°$ and an effective aperture of $a_e^u + a_e^l \cong 918\,\mu m$.

9.5 Selfoc Lens

Within the communication area, most optical devices for manipulating and processing signals in optical fiber transmission systems include lenses. For most of these devices GRIN lenses have a number of advantages over conventional lenses, as discussed in Sect. 9.2–9.4, such as small size, light weight, low cost, short focal distances, easy mounting and adjustment, and stable configuration. A GRIN lens carries out such typical functions as on–axis imaging, collimation, focusing, and off–axis imaging, which are of great importance in optical communications systems. On the one hand, because of the relatively large numerical aperture of optical fibers, in fiber devices lenses are needed to convert the diverging beams from the input fibers into converging beams that can couple efficiently to the output fibers. On the other hand, for use with angularly sensitive elements such as interference filters or gratings, fiber devices need lenses to convert the diverging beams from the input fibers into collimated beams for processing the light by the angularly sensitive elements, and then to transform the resulting collimated output beams into converging beams that are coupled to the output fibers. Taking advantage of the inherent functions of the GRIN lenses, devices that use selfoc lenses for optical fiber transmission systems have been developed and reported [9.19–9.20].

Figure 9.8 shows selfoc collimator and focuser devices and how the small–diameter, large numerical aperture beam from the input fiber is transformed by a quarter–pitch selfoc lens into large–diameter collimated beam, or how a collimated beam can be focused in the output fiber. In case (a), the lens performs a spatial Fourier transform on the input beam, and at the output face of the lens a collimated beam is obtained. In case (b), the quarter–pitch selfoc lens converts the collimated beam into an image of the input beam that is coupled to the output fiber. Figure 9.8 shows the basic optical configuration used to form a collimated beam in free space using a pair of selfoc lenses. Free space between collimator and focuser lenses allows one, for instance, to insert or to remove at will attenuator elements. An attenuator is mainly used to equalize optical signals. The total attenuation of the device is the sum of the attenuations of the individual elements, independent of their order.

A connector is the simplest application of selfoc lenses for fiber–fiber coupling. As an example, Fig. 9.9 shows a numerical aperture conversion using a selfoc lens in optical fiber lines. The configuration is based on the on–axis imaging properties of the selfoc lens. The light beam coming from a fiber is coupled to another fiber by a selfoc lens. The converted beam must have a numerical aperture that is close to the numerical aperture of the output fiber. Then, the selfoc lens converts the numerical aperture of the beam. The butt–joint connector, in which the two fibers are simple butted against each other, is a viable alternative since it does not use any lenses. In a butt–joint connector the fiber must be positioned with high precision, but the lens connector is relatively insensitive to linear misalignments. Likewise, an air gap can be tolerated between the connector plugs. The selfoc lens connector may be advantageous for use in dusty environments and when connectors must be disconnected and reconnected many times.

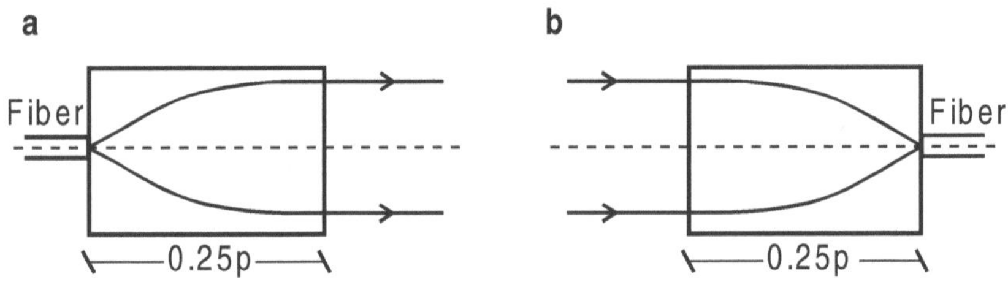

Fig. 9.8. A quarter–pitch selfoc lens **a** collimates or **b** focuses incoming light.

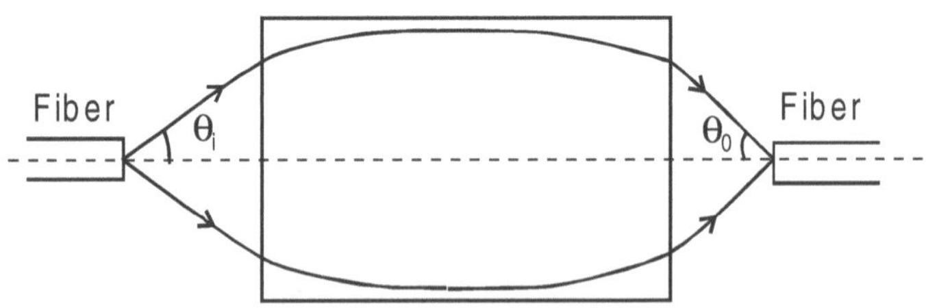

Fig. 9.9. Numerical aperture conversion using a selfoc lens.

The problem of dividing the beam from a fiber into two or more beams and coupling those beams into two or more output fibers can be solved by a directional coupler. Figure 9.10 illustrates a four–port directional coupler using two selfoc lenses. The device uses a beam–splitter film between two selfoc lenses of quarter–pitch arranged in series. The optical configuration is based on off–axis imaging properties of the GRIN lenses, so that each lens serves two fibers, and only two lenses are required. If the beam–splitter film in a directional coupler is chosen to have a reflectivity and transmissivity that vary significantly with the wavelength of the incident light, the coupler becomes a multiplexer or demultiplexer. A wavelength–division de/multiplexer using an interference filter is shown in Fig.9.11. This de/multiplexer is a typical device that demonstrates the position–angle transformation feature of the selfoc lenses. Since a single filter can only divide an optical beam into two output beams, to separate m channels requires at least m–1 separate filters. A multichannel de/multiplexer can be made by using fibers to connect a number of basic devices of the type illustrated in Fig. 9.10.

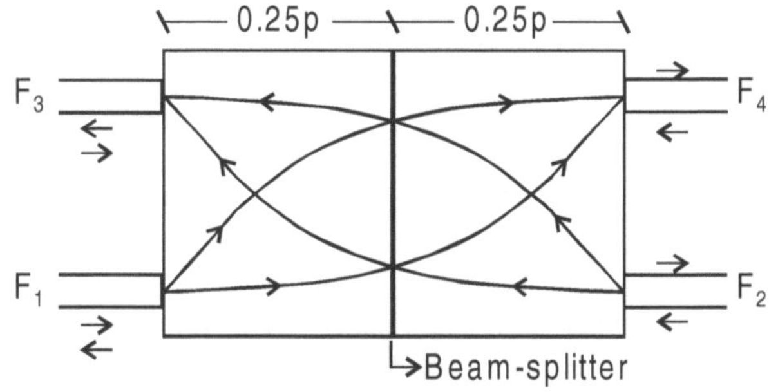

Fig. 9.10. A four–port directional coupler using two selfoc lenses.

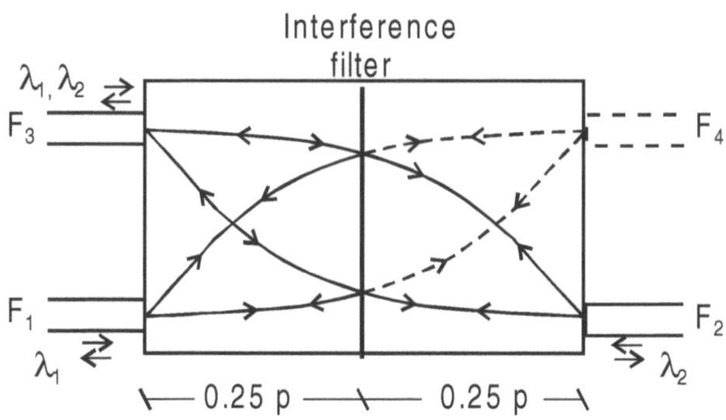

Fig. 9.11. A wavelength–division de/multiplexer using two selfoc lenses.

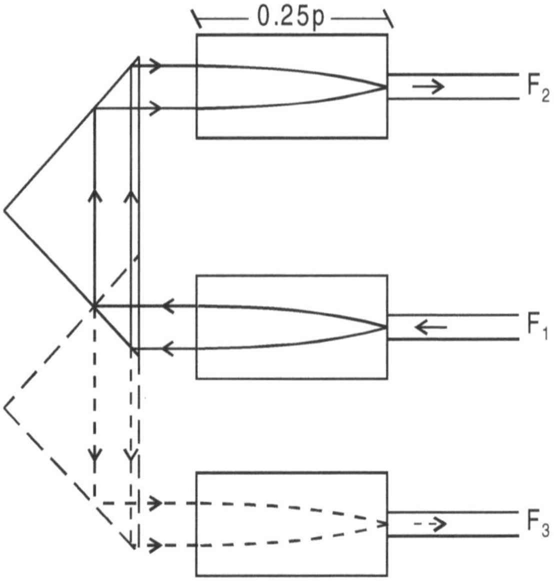

Fig. 9.12. Optical switch using selfoc lenses and a movable prism.

An optical switch is one of the key devices in optical fiber transmission systems, and many kinds of switches use selfoc lenses with mirrors or prisms. Figure 9.12 shows a two–position switch that has three quarter–pitch selfoc lenses and a movable prism to translate the collimated beam. With the prism in the upper position, light is coupled between fibers F_2 and F_1. When the prism is shifted to the lower position, light is coupled between fibers F_3 and F_1. The switch works with a collimated beam, and the moving element need not be positioned with great precision. Multiposition switches can be made using the same principle as the two–position switches by arranging the moving element to couple the input beam between a common lens and any one of a set of other lenses. Figure 9.13 illustrates a multichannel optical switch using a rotating mirror. The common fiber is centered on the selfoc lens axis, and the other fibers are arranged in a circular array around it on the same lens. A plane mirror mounted on a mechanical rotator produces a beam deflection such that the light is coupled between the common fiber and any one of the fibers in the circular array.

An optical bus interconnection system consisting of cascade arrays of selfoc lenses is used as a free–space three–dimensional optical interconnect dealing with a large amount of information in a complicated interconnection network [9.21–9.23]. Figure 9.14 depicts the configuration of the system, which is applied to board–to–board interconnections. Selfoc lenses are aligned and fixed on a glass substrate with grooves, then perpendicular gaps are fabricated by using a slicing machine. The position of each individual gap is properly determined so that the conjugate image planes of unit magnification are located at the same position in all gaps. In practice, a selfoc rod is divided into many collimated lenses, and they form a telecentric optical system that is suitable for cascade interconnections. It also permits the possibility of using the Fourier planes. When a matrix array of LEDs or laser diodes is placed at one end of the selfoc rod, the image of the light pattern can be transmitted to all the conjugate planes. A signal generated at the light source can be led into many electronic circuit boards if boards with transparent photodetector arrays are inserted into the gaps.

Selfoc lenses can also be used for optical sensing [9.24]. Intensity–modulated fiber–optic sensors have been described in the literature for the measurement of a variety of environmental parameters such as pressure, fluid flow, temperature, and acceleration [9.25–9.26]. Many of these sensors require the direct measurement of a physical displacement that is used to infer the parameter of interest and employ GRIN lenses or GRIN fiber lenses in order to improve the sensing characteristics.

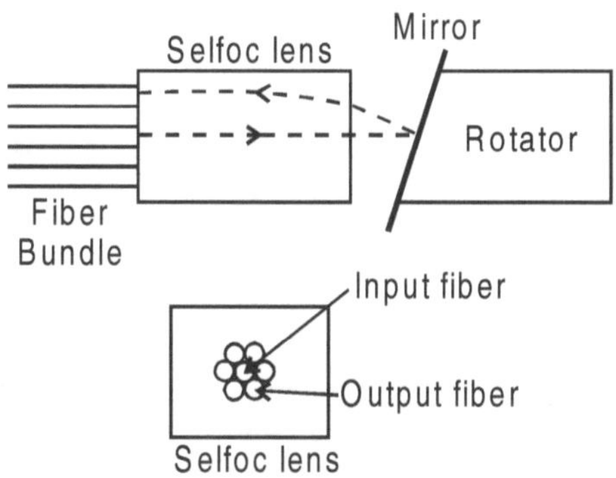

Fig. 9.13. Multichannel optical switch.

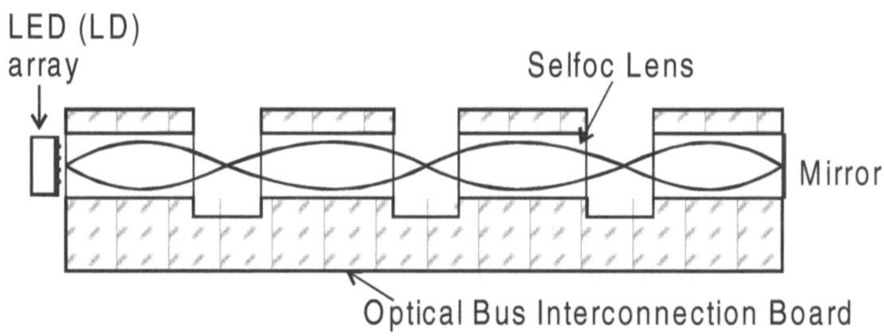

Fig. 9.14. Optical bus interconnection system using selfoc lenses.

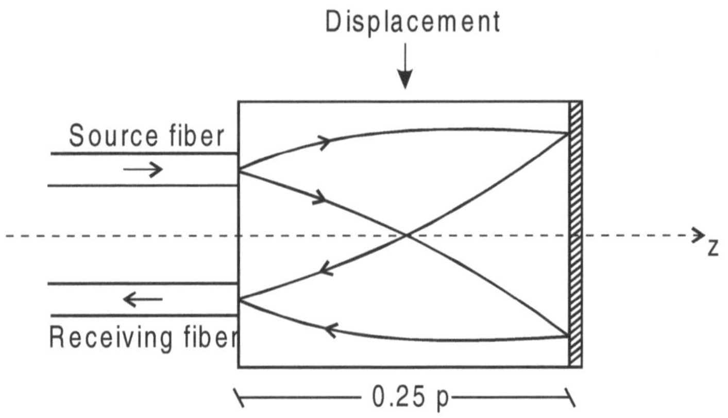

Fig. 9.15. Fiber–optic lateral displacement sensor using a quarter–pitch selfoc lens.

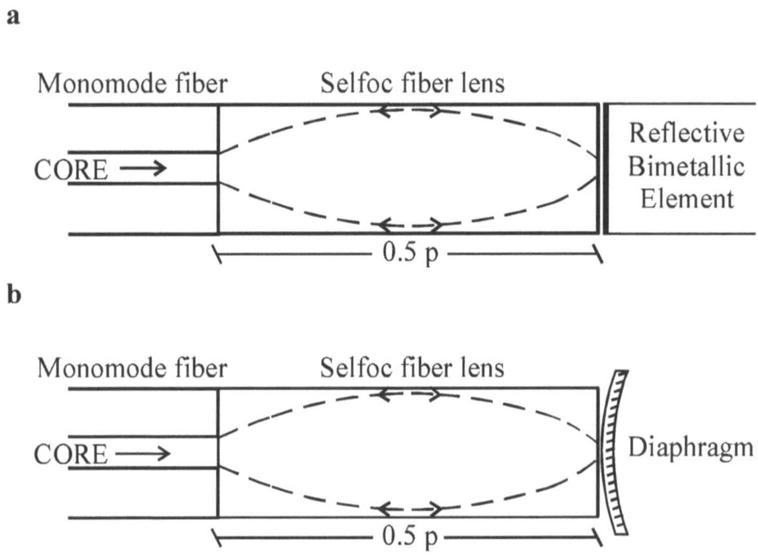

Fig. 9.16. Fiber–optic temperature **a** and pressure **b** sensors using a half–pitch selfoc fiber lens.

An important example of this kind of sensors has been reported [9.27]. The optical sensor consists of a pair of optical fibers (dual–fiber configuration) that are adjacent to a quarter–pitch selfoc lens having a reflective coating at one end. The two fibers are maintained at a fixed position relative to each other, symmetrically located with respect to the lens axis. At this position, optimum coupling efficiency is achieved. The sensor can be visualized as in Fig. 9.10, when the second quarter–pitch selfoc lens is removed and the film is a totally reflecting coating (Fig. 9.15).

Light from the source fiber is directed into the quarter–pitch selfoc lens, reflected by the mirrored back surface, and collected by the receiving fiber. The mirrored surface of the lens produce a folded half–pitch selfoc lens, expanding the light from the source fiber, reversing its direction, and refocusing the light in the receiving fiber. Intensity modulation results from the relative transverse displacement between the lens and the pair of fibers. The transverse displacement technique can also be used with a single optical fiber (single–fiber configuration) located along the lens axis that acts as both source and receiver by employing a beam splitter between the light source and the fiber.

Many variants of this sensing structure can be implemented, when a monomode fiber is fused and aligned to a half–pitch selfoc fiber lens, as illustrated in Fig.9.16. This single–fiber configuration can also be visualized as in Fig. 9.1, when the second monomode fiber is replaced by a target located in front of the selfoc fiber lens such that its position or state is a function of some physical parameter to be sensed. Targets such as a bimetallic element (for temperature sensing) or a flexible diaphragm with a reflective inner surface (for pressure sensing), which varying their distances from the selfoc fiber lens in response to temperature or pressure inputs, can be used for an intensity modulation sensor as is shown in Fig. 9.16. The single–mode fiber is connected to the target by the selfoc fiber lens. The light from the source fiber is focused on the output face of the half–pitch selfoc fiber lens, and the light reflected by the mirrored surface of the target is coupled into the fiber. The coupling efficiency is very sensitive to axial displacement and can be evaluated by using a Gaussian mode approximation, as discussed in Sect. 9.2, in such a way that the design of the sensor based on optical fibers and GRIN lenses or GRIN fiber lenses is greatly simplified.

The most attractive advantage of using GRIN lenses or GRIN fiber lenses in a intensity–modulated sensor is that they can be used to increase sensor sensitivity by improving the coupling efficiency by reducing misalignments and, therefore, producing a stable sensing structure. This advantage has been taken into account in a sapphire single–crystal optical high–temperature thermometer [9.28]. GRIN lenses were applied to improve the coupling efficiency when there is a greater error induced by aligning or machining the optical fiber connector between the sensing head and the detector.

Finally, one of the important issues in color image sensor development has been to reproduce colors accurately. However, considerable color distortion occurs, especially in picture regions such as pattern edges, where brightness or color changes drastically. Spurious coloration of the boundaries around pattern edges produces inaccurate color reproduction. A color image sensor with a GRIN rod lens has been designed to quantify the relationship between color aberration and color distortion around pattern edges [9.29].

References

Preface

P.1 R.K. Luneburg: *Mathematical Theory of Optics* (University of California Press, Berkeley, 1964) and *Mimeographed Lectured Notes on Mathematical Theory of Optics* (Brown University, Providence 1944)
P.2 E.W. Marchand: *Gradient Index Optics* (Academic Press, New York 1978)
P.3 Y. LeGrand, S.G. ElHage: *Physiological Optics* (Springer-Verlag, Berlin 1980)
P.4 D.A. Atchison, G. Smith: *Optics of the Human Eye* (Butterworth-Heinemann, Melbourne 2000)
P.5 J.M. Enoch, F.L. Tobey, Jr., eds.: *Vertebrate Photoreceptor Optics* (Springer-Verlag, Berlin 1981)
P.6 A.B. Fraser, W.H. Mach: Sci. Am. **234**, 102 (1976)
P.7 S.N. Houde-Walter, D.T. Moore: Appl. Opt. **25**, 3373 (1986)
P.8 D.T. Moore: Appl. Opt. **19**, 1035 (1980)
P.9 S.N. Houde-Walter: Laser Focus World **25**, 151 (1989)
P.10 Y. Koike: OSA Annual Meeting, Technical Digest Series, 261 (1992)
P.11 J.B. McChesney: Proc. IEEE **68**, 1181 (1980)
P.12 R. Blankenbecler, G.E. Rindone: OSA Annual Meeting, Technical Digest Series, 253 (1992)
P.13 J.C. Maxwell: Cambridge and Dublin Math. J. **8**, 88 (1854)
P.14 R.W. Wood: *Physical Optics* (McMillan, New York 1905)
P.15 R. Stettler: Optik **12**, 529 (1955)
P.16 S.P. Morgan: J. Appl. Phys. **29**, 1358 (1958)
P.17 J. Sochacki: J. Opt. Soc. Am. **73**, 789 and 1839 (1983)
P.18 J. Sochacki: Appl. Opt. **23**, 4444 (1984)
P.19 F. Zernike: Opt. Commun. **12**, 579 (1974)
P.20 W.H. Southwell: J. Opt. Soc. Am. **67**, 1004 and 1010 (1977)
P.21 J. Sochacki: Opt. Commun. **41**, 13 (1981)
P.22 E. Colombini: J. Opt. Soc. Am. **71**, 1403 (1981)
P.23 S. Doric, E. Munro: J. Opt. Soc. Am. **73**, 1083 (1983)
P.24 J. Sochacki: J. Mod. Opt. **35**, 891 (1988)
P.25 A.L. Mikaelian: SU Academy Reports **81**, 569 (1951), in Russian
P.26 A. Fletcher, T. Murphy, A. Young: Proc. R. Soc. London, Ser. A **223**, 216 (1954)
P.27 Y. Silberberg, U. Levy: J. Opt. Soc. Am. **69**, 960 (1979)
P.28 T. Tamir, ed.: *Guided-Wave Optoelectronics* (Springer-Verlag, Berlin 1990)
P.29 C. Paré, L. Gagnon, P.A. Bélanger: Phys. Rev. A **46**, 4150 (1992)
P.30 A.O. Barut, A. Inomata, R. Wilson: J. Phys. A **20**, 4083 (1987)
P.31 L. Montagnino: J. Opt. Soc. Am. **58**, 1667 (1968)
P.32 F.P. Kapron: J. Opt. Soc. Am. **60**, 1433 (1970)
P.33 P.J. Sands: J. Opt. Soc. Am. **60**, 1436 (1970)

P.34 P.J. Sands: J. Opt. Soc. Am. **61**, 777, 879, 1086 and 1495 (1971)
P.35 P.J. Sands: J. Opt. Soc. Am. **63**, 1210 (1973)
P.36 D.T. Moore: J. Opt. Soc. Am. **61**, 886 (1971)
P.37 D.T. Moore, P.J. Sands: J. Opt. Soc. Am. **61**, 1195 (1971)
P.38 A. Gupta, K. Thyagarajan, I.C. Goyal, A.K. Ghatak: J. Opt. Soc. Am. **66**, 1320 (1976)
P.39 K.B. Paxton, W. Streifer: Appl. Opt. **10**, 769 and 2090 (1971)
P.40 J. Brushon: Opt. Acta **22**, 211 (1975)
P.41 M.E. Harrigan: Appl. Opt. **23**, 2702 (1984)
P.42 M.S. Sodha, A.K. Ghatak: *Inhomogeneous Optical Waveguides* (Plenum Press, New York 1977)
P.43 Y.A. Kravtsov, Y.I. Orlov: *Geometrical Optics of Inhomogeneous Media* (Springer-Verlag, Berlin 1990)
P.44 G.I. Greisukh, S.T. Bobrov, S.A. Stepanov: *Optics of Diffractive and Gradient-Index Elements and Systems* (SPIE Optical Engineering Pres, Bellingham 1997)
P.45 A. Yariv: J. Opt. Soc. Am. **66**, 301 (1976)
P.46 K. Iga, Y. Kokubun, M. Oikawa: *Fundamentals of Microoptics* (Academic Press, New York 1984)
P.47 C. Gomez-Reino, E. Larrea: Appl. Opt. **21**, 4271 (1982)
P.48 C. Gomez-Reino, M.V. Perez, E. Larrea: Opt. Commun. **44**, 8 (1982)
P.49 C. Gomez-Reino, E. Larrea: Appl. Opt. **22**, 387 (1983)
P.50 I. Kitano, K. Koizumi, H. Matsumura, T. Uchida, M. Furukawa: Jpn. J. Appl. Phys. **39**, 63 (1970)
P.51 M. Kawazu, Y. Ogura: Appl. Opt. **19**, 1105 (1980)
P.52 J.D. Rees, W. Lama: Appl. Opt. **23**, 1715 (1984)
P.53 J.F. Goldenberg, J.J. Miceli, D.T. Moore: Appl. Opt. **24**, 4288 (1985)
P.54 I.P. Csorba: Appl. Opt. **19**, 1139 (1980)
P.55 T. Uchida, M. Furukawa, I. Kitano, H. Matsumura: J. Quant. Electron. QE-**6**, 606 (1970)
P.56 G. Von Bally, W. Schmidthaus, H. Sakowski, W. Mette: Appl. Opt. **23**, 1725 (1984)
P.57 G. Von Bally, E. Brune, W. Mette: Appl. Opt. **25**, 3425 (1986)
P.58 D. Gloge, E.A.J. Marcatili: Bell. Syst. Tech. J. **52**, 1563 (1973)
P.59 K. Kikuchi, S. Ishihara, H. Shimuzi, J. Shimada: Appl. Opt. **19**, 1076 (1980)
P.60 M. Daimon, M. Shinoda, T. Kubo: Appl. Opt. **23**, 1790 (1984)
P.61 H. Nishi, H. Ichikawa, M. Toyama, I. Kitano: Appl. Opt. **25**, 3340 (1986)
P.62 K. Kobayashi, R. Ishikawa, K. Minemura, S. Sugimoto: Fiber and Integrated Opt. **2**, 1 (1979)
P.63 W.J. Tomlinson: Appl. Opt. **19**, 1127 (1980)
P.64 K. Hamanaka: Opt. Lett. **16**, 1222 (1991)
P.65 P.J. Murphy, T.P. Coursolle: Appl. Opt. **29**, 544 (1990)
P.66 W.H. Hu, C.F. Wang, W.Z. Tang, W. Zhou, J.H. Zhou: Int. J. Optoelectron. **7**, 121 (1992)
P.67 P.K.C. Pillai, T.C. Goel, N.K. Pandey, S.K. Nijhawan: Int. J. Optoelectron. **7**, 429 (1992)

Chapter 1

1.1 M. Born, E. Wolf: *Principles of Optics* (Pergamon Press, New York 1975), Chap. 3
1.2 R.K. Luneburg: *Mathematical Theory of Optics* (University of California Press, Berkeley 1964), Chaps. 1–3
1.3 D. Marcuse: *Light Transmission Optics*, 2nd edn. (Van Nostrand Reinhold, New York 1982), Chap. 1
1.4 J.A. Arnaud: *Beam and Fiber Optics* (Academic Press, New York 1976), Chap. 3

1.5 E.W. Marchand: *Gradient Index Optics* (Academic Press, New York 1978), Chap. 1
1.6 D.T. Moore: J. Opt. Soc. Am. **65**, 451 (1975)
1.7 M.S. Sodha, A.K. Ghatak: *Inhomogeneous Optical Waveguides* (Plenum Press, New York 1977), Chap. 9
1.8 I.S. Gradshteyn, I.M. Ryzhik: *Tables of Integrals, Series and Products* (Academic Press, New York 1980), Chap. 16
1.9 C. Gomez-Reino, J. Sochacki: Appl. Opt. **24**, 4375 (1985)
1.10 M. Sochacka, J. Sochacki, C. Gomez-Reino: Appl. Opt. **25**, 142 (1986)
1.11 M.E. Harrigan: Appl. Opt. **23**, 2702 (1984)

Chapter 2

2.1 R.K. Luneburg: *Mathematical Theory of Optics* (University of California Press, Berkeley 1964), Chap. 4
2.2 M. Nazarathy, J. Shamir: J. Opt. Soc. Am. **72**, 356 (1982)
2.3 M. Nazarathy, J. Shamir: J. Opt. Soc. Am. **72**, 1398 (1982)
2.4 S. Abe, J.T. Sheridan: J. Phys. A **27**, 4179 (1994)
2.5 S. Abe, J.T. Sheridan: Opt. Lett. **19**, 1801 (1994)
2.6 S. Abe, J.T. Sheridan: Opt. Commun. **113**, 385 (1994)
2.7 A.W. Lohmann: J. Opt. Soc. Am. A **10**, 2181 (1993)
2.8 T. Alieva, F. Agullo-Lopez: Opt. Commun. **114**, 161 (1995)
2.9 C. Gomez-Reino: Int. J. Optoelectron. **7**, 607 (1992)
2.10 T. Alieva: *Application of the Canonical Integral Transform to Optical Propagation Problems*, Ph.D. Dissertation, Autonomous University of Madrid (1996)
2.11 G. Eichmann: J. Opt. Soc. Am. **61**, 161 (1971)
2.12 C. Gomez-Reino, E. Larrea: Appl. Opt. **21**, 4271 (1982)
2.13 C. Gomez-Reino, E. Larrea: Appl. Opt. **22**, 387 (1983)
2.14 C. Gomez-Reino, M.V. Pérez, E. Larrea: Opt. Commun. **44**, 8 (1982)
2.15 V. Namias: J. Inst. Math. Appl. **25**, 241 (1980)
2.16 A.C. McBride, F.H. Kerr: J. Appl. Math. **39**, 159 (1987)
2.17 D. Mendlovic, H.M. Ozaktas: J. Opt. Soc. Am. A **10**, 1875 (1993)
2.18 H.M. Ozaktas, D. Mendlovic: J. Opt. Soc. Am. A **10**, 2522 (1993)
2.19 D. Mendlovic, H.M. Ozaktas: Opt. Commun. **101**, 163 (1993)
2.20 D. Mendlovic, H.M. Ozaktas, A.W. Lohmann: Appl. Opt. **33**, 6188 (1994)
2.21 D. Mendlovic, Y. Bitran, R.G. Dorsch, C. Ferreira, J. Garcia, H.M. Ozaktas: Appl. Opt. **34**, 7451 (1995)
2.22 J. Garcia, D. Mendlovic, Z. Zalevsky, A. Lohmann: Appl. Opt. **35**, 3945 (1996)
2.23 A.W. Lohmann, D. Mendlovic, Z. Zalevsky: `Fractional Transformations in Optics´. In *Progress in Optics XXXVIII*, E. Wolf, ed. (Elsevier Science B.V., Amsterdam 1998) pp. 263–342
2.24 C. Gomez-Reino, M.V. Perez, E. Larrea: Opt. Commun. **45**, 372 (1983)
2.25 K.E. Oughstun: `Unstable Resonator Modes´. In *Progress in Optics XXIV*, E. Wolf, ed. (Elsevier Science B.V., Amsterdam 1987) pp. 165–387
2.26 A.Yariv: J. Opt. Soc. Am. **66**, 302 (1976)

Chapter 3

3.1 *Selfoc Catalog* (Nippon Sheet Glass Co. Ltd., Tokyo 1987)

3.2 C. Gomez-Reino, J. Liñares, E. Larrea: J. Opt. Soc. Am. A **3**, 1604 (1986)
3.3 C. Gomez-Reino, J. Liñares, J. Sochacki: Appl. Opt. **25**, 1076 (1986)
3.4 C. Gomez-Reino, J. Liñares: J. Opt. Soc. Am. A **4**, 1337 (1987)
3.5 C. Gomez-Reino, J. Liñares: Appl. Opt. **25**, 3418 (1986)
3.6 J.W. Goodman: *Introduction to Fourier Optics*, 2nd edn. (McGraw-Hill, New York 1995), Chap. 5
3.7 A. Vanderlugt: *Optical Signal Processing* (John Wiley and Sons, New York 1992), Chap. 3
3.8 J.R. Flores, C. Gomez-Reino, E. Acosta, J. Liñares: Opt. Eng. **28**, 1173 (1989)
3.9 L. Montagnino: J. Opt. Soc Am. **58**, 1667 (1968)
3.10 E.W. Marchand: J. Opt. Soc. Am. **60**, 1 (1970) and Appl. Opt. **11**, 1104 (1972)
3.11 F.P. Kapron: J. Opt. Soc. Am. **60**, 1433 (1970)
3.12 P.J. Sands: J. Opt. Soc. Am. **61**, 879 (1971)
3.13 W. Streifer, K.P. Paxton: Appl. Opt. **20**, 769 (1971)
3.14 D.T. Moore: J. Opt. Soc. Am. **65**, 451 (1975)
3.15 S.J.S. Brown: Appl. Opt. **19**, 1056 (1980)
3.16 M.E. Harrigan: Appl. Opt. **23**, 2702 (1984)
3.17 E. Acosta, C. Gomez-Reino, J. Liñares: Appl. Opt. **26**, 2952 (1987)
3.18 J. Rogers, M.E. Harrigan, R.P. Loce: Appl. Opt. **27**, 452 and 459 (1988)
3.19 C. Gomez-Reino, E. Acosta, M.V. Perez: Jpn. J. Appl. Phys. **31**, 1582 (1992)
3.20 C. Bao, C. Gomez-Reino, M.V. Perez: Opt. Commun. **172**, 1 (1999)
3.21 K. Iga, Y. Kokubun, M. Oikawa: *Fundamentals of Microoptics* (Academic Press, New York 1984), Chap. 5
3.22 C. Gomez-Reino, E. Larrea: Appl. Opt. **22**, 970 (1983)
3.23 C. Gomez-Reino, M.V. Perez, E. Larrea: Opt. Commun. **47**, 369 (1983)
3.24 I.S. Gradshteyn, I.M. Ryzhik: *Tables of Integrals, Series and Products* (Academic Press, New York 1980), Sect. 8.57
3.25 M. Born, E. Wolf: *Principles of Optics* (Pergamon Press, New York 1975), Chap. 8
3.26 C. Gomez-Reino, E. Acosta, J. Liñares: J. Mod. Opt. **34**, 1501 (1987)

Chapter 4

4.1 B.E.A. Saleh, M.C. Teich: *Fundamentals of Photonics* (John Wiley and Sons, New York 1991), Chap. 3
4.2 J.A. Arnaud: *Beam and Fiber Optics* (Academic Press, New York 1976), Chap. 2
4.3 D. Marcuse: *Light Transmission Optics*, 2nd edn. (Van Nostrand Reinhold, New York 1982), Chap. 6
4.4 A. Siegman: *Laser* (University Science, Mill Valley 1986), Chaps. 15–17.
4.5 W.T. Silfvast: *Laser Fundamentals* (Cambridge University Press, New York, 1996), Chapter 11.
4.6 H. Kogelnik: Appl. Opt. **4**, 1562 (1965)
4.7 P. A. Belanger: Opt. Lett. **16**, 196 (1991)
4.8 L.W. Casperson: Appl. Opt. **12**, 2434 (1973)
4.9 L.W. Casperson: Appl. Opt. **20**, 2243 (1981)
4.10 L.W. Casperson: Appl. Opt. **24**, 4395 (1985)
4.11 L.W. Casperson, J.L. Kirkwood: J. Lightwave Technol. **3**, 264 (1985)
4.12 J.N. McMullin: Appl. Opt. **25**, 2184 (1986)
4.13 J.N. McMullin: Appl. Opt. **28**, 1298 (1989)
4.14 A.A. Tovar, L.W. Casperson: Appl. Opt. **33**, 7733 (1994)
4.15 C. Gomez-Reino: Int. J. Optoelectron. **7**, 607 (1992)

4.16 E. Acosta, J.R. Flores, C. Gomez-Reino, J. Liñares: Opt. Eng. **28**, 1168 (1989)
4.17 W.H. Carter: Appl. Opt. **21**, 1989 (1982)
4.18 S.A. Self: Appl. Opt. **22**, 658 (1983)
4.19 H. Weber: Opt. Commun. **62**, 124 (1987)
4.20 *Selfoc Catalog* (Nippon Sheet Glass Co. Ltd., Tokyo 1987)
4.21 C. Gomez-Reino, M.V. Perez, J. Sochacki, M. Sochacka: Opt. Commun. **55**, 5 (1985)
4.22 J. Liñares, C. Gomez-Reino: J. Mod. Opt. **38**, 481 (1991)

Chapter 5

5.1 A. Yariv: *Quantum Electronics*, 2nd edn. (Wiley and Sons, New york 1975), Chap. 6
5.2 J.A. Arnaud: *Beam and Fiber Optics* (Academic Press, New York 1976), Chaps. 2, 4
5.3 D. Marcuse: *Light Transmission Optics*, 2nd edn. (Van Nostrand Reinhold, New York 1982), Chap. 7
5.4 A.E. Siegman: *Lasers* (University Science, Mill Valley 1986), Chap. 20
5.5 N. Hodgson, H. Weber: *Optical Resonators: Fundamentals, Advanced Concepts and Applications* (Springer-Verlag, Berlin 1997), Part IV
5.6 L.W. Casperson: J. Opt. Soc. Am. **62**, 1373 (1976)
5.7 L.W. Casperson: J. Lightwave Technol. **3**, 264 (1985)
5.8 A.A. Tovar, L.W. Casperson: IEEE Trans. Microwave Theory Tech. **43**, 1857 (1995)
5.9 L.W. Casperson, D.G. Hall, A.A. Tovar: J. Opt. Soc. Am. A **14**, 3341 (1997)
5.10 L.W. Casperson, A.A. Tovar: J. Opt. Soc. Am. A **15**, 954 (1998)
5.11 A. E. Siegman: J. Opt. Soc. Am. **63**, 1093 (1973)
5.12 L.W. Casperson, S.D. Lunnam: Appl. Opt. **14**, 1193 (1975)
5.13 M.E. Smithers: Appl. Opt. **25**, 118 (1986)
5.14 A.A. Tovar, L.W. Casperson: J. Opt. Soc. Am. A **8**, 6 (1991)
5.15 A.A. Tovar, L.W. Casperson: J. Opt. Soc. Am. A **12**, 1522 (1995)
5.16 A.A. Tovar, L.W. Casperson: J. Opt. Soc. Am. A **13**, 90 (1996)
5.17 M. Nazarathy, J. Shamir: J. Opt. Soc. Am. **72**, 1398 (1982)
5.18 M. Nazarathy, A. Hardy, J. Shamir: J. Opt. Soc. Am. **72**, 1409 (1982)
5.19 J. Liñares, C. Gomez-Reino, J.R. Flores, E. Acosta: J. Mod. Opt. **35**, 679 (1988)
5.20 Y. Li: J. Opt. Soc. Am. A **3**, 1761 (1986)
5.21 Y. Li, E. Wolf: Opt. Commun. **42**, 151 (1982)
5.22 H. Kogelnik: Appl. Opt. **4**, 1562 (1965)
5.23 J.B. Keller, W. Streifer: J. Opt. Soc. Am. **61**, 40 (1971)
5.24 J.A. Arnaud: Appl. Opt. **24**, 538 (1985)
5.25 A. Kujawski: Appl. Opt. **28**, 2458 (1989)
5.26 R. Martinez-Herrero, P.M. Mejias: Opt. Commun. **85**, 162 (1991)
5.27 C. Gomez-Reino, E. Acosta, R. Gonzalez, J. Liñares, J.R. Flores: J. Mod. Opt. **38**, 317 (1991)

Chapter 6

6.1 K. Iga, Y. Kokubun, M. Oikawa: *Fundamentals of Micro–optics* (Academic Press, Tokyo 1984), Chap. 1
6.2 R.K. Luneburg: *Mathematical Theory of Optics* (University of California Press, Berkeley 1964), Chaps. 3–4
6.3 E.W. Marchand: *Gradient Index Optics* (Academic Press, New York 1978), Chap. 5

6.4 M.S. Sodha, A.K. Ghatak: *Inhomogeneous Optical Waveguides* (Plenum Press, New York 1977), Chap. 9
6.5 A. Fletcher, T. Murphy, A. Young: Proc. R. Soc. London, Ser. A **223**, 216 (1954)
6.6 D.W. Hewak, J.W.Y. Lit: Appl. Opt. **28**, 4190 (1989)
6.7 R.V. Ramaswamy, R. Srivastava: J. Lightwave Technol. **6**, 984 (1988)
6.8 N. Takoto, K. Jinguji, M. Yasu, H. Toba, M. Kawachi: J. Lightwave Technol. **6**, 1003 (1988)
6.9 C. Bao, C. Gomez-Reino: Opt. Quantum. Electron. **27**, 897 (1995)
6.10 C. Bao, M.V. Perez, C. Gomez-Reino: Opt. Lett. **21**, 1078 (1996)
6.11 C. Bao, C. Gomez-Reino, M.V. Perez: J. Opt. Soc. Am. A **14**, 1754 (1997)
6.12 M.V. Perez, C. Bao, C. Gomez-Reino: Opt. Commun. **146**, 302 (1998)
6.13 C. Gomez-Reino, C. Bao, M.V. Perez, Y. Wang: J. Mod. Opt. **45**, 1785 (1998)
6.14 M.V. Perez, C. Bao, C. Gomez-Reino: Jpn. J. Appl. Phys. **37**, 3638 (1998)
6.15 T. Tamir, ed.: *Guided-Wave Optoelectronics* (Springer-Verlag, Berlin 1990), Chap. 2
6.16 M.J. Adams: *Introduction to Optical Waveguides* (Wiley, New York 1981), Chap. 4
6.17 A.W. Snyder, J.D. Love: *Optical Waveguide Theory* (Chapman and Hall, London 1983), Chap. 12
6.18 T. Kozek: Proc. SPIE **670**, 226 (1986) and Proc SPIE **704**, 44 (1986)
6.19 C. Paré, L. Gagnon, P.A. Bélanger: Phys. Rev. A **46**, 4150 (1992)
6.20 I.S. Gradshteyn, I.M. Ryzhik: *Tables of Integrals, Series and Products* (Academic Press, New York 1980), Sect. 9.1
6.21 J. Liñares, C. Gomez-Reino: Appl. Opt. **33**, 3427 (1994)
6.22 A. Papoulis: *Systems and Transforms with Applications in Optics* (Prentice-Hall, Englewood Cliffs 1968), Chap. 7
6.23 J.J. Stammes: *Waves in Focal Regions* (Adam-Hilger, Bristol 1986), Chap. 8
6.24 C. Gomez-Reino, M.V. Perez, C. Bao, M.T. Flores-Arias, S. Vidal, S. Fernandez de Avila: Jpn. J. Appl. Phys. **39**, 1463 (2000)
6.25 C. Gomez-Reino, M.V. Perez, C. Bao, M.T. Flores-Arias, S. Vidal, S. Fernandez de Avila: Appl. Opt. **39**, 2145 (2000)
6.26 C. Gomez-Reino, M.V. Perez, C. Bao, M.T. Flores-Arias, S. Vidal: J. Mod. Opt. **47**, 91 (2000)

Chapter 7

7.1 H.F. Talbot: Phil. Mag. **9**, 401 (1836)
7.2 Lord Rayleigh: Phil. Mag. **11**, 196 (1881)
7.3 W.D. Montgomery: J. Opt. Soc. Am. **57**, 772 (1967)
7.4 K. Patorski: 'The Self-Imaging Phenomenon and Its Application'. In *Progress in Optics XXVII*, E. Wof, ed. (Elsevier Science B.V., Amsterdam 1989) pp. 3–101
7.5 P.M. Mejias, R. Martinez: J. Opt. Soc. Am. A **8**, 266 (1991)
7.6 M.V. Berry, S. Klein: J. Mod. Opt. **43**, 2139 (1996)
7.7 J.R. Leger, G.J. Swanson: Opt. Lett. **15**, 288 (1990)
7.8 A.W. Lohmann, J.A. Thomas: Appl. Opt. **29**, 4337 (1990)
7.9 V. Arrizon, J. Ojeda-Castañeda: Opt. Lett. **18**, 1 (1993)
7.10 C. Zhou, L. Wang, T. Tschudi: Opt. Commun. **147**, 224 (1998)
7.11 J.F. Clauser, S. Li: Phys. Rev. A **49**, 2213 (1994)
7.12 J.F. Clauser, M.W. Reinsch: Appl. Phys. B **54**, 380 (1992)
7.13 P. Szwaykowski, J. Ojeda-Castañeda: Opt. Commun. **83**, 1 (1991)

7.14 R.G. Hunsperger: *Integrated Optics: Theory and Technology*, 4th edn. (Springer-Verlag, Berlin 1995), Chap. 6
7.15 Th. Niemeier, R. Ulrich: Opt. Lett. **11**, 677 (1986)
7.16 W-H Yeh, M. Mansuripur, M. Fallahi, R. Scott: Opt. Commun. **170**, 207 (1999)
7.17 C. Aramburu: *Photonic Integrated Devices by Multimode Interference for Optical Networks*, Ph.D. Dissertation, Telecommunication High School, Public University of Navarra (2000)
7.18 M.T. Flores-Arias, C. Bao, M.V. Perez, C. Gomez-Reino: J. Opt. Soc. Am. A **16**, 2439 (1999)
7.19 R. Jozwicki: Opt. Acta **30**, 73 (1983)
7.20 G.S. Agarwal: Opt. Commun. **119**, 30 (1995)
7.21 D. Joyeux, Y. Cohen-Sabban: Appl. Opt. **21**, 625 (1982)
7.22 Y. Cohen-Sabban, D. Joyeux: J. Opt. Soc. Am. **73**, 707 (1983)
7.23 V. Arrizon, J.G. Ibarra: Opt. Lett. **21**, 378 (1996)
7.24 V. Arrizon, J. Ojeda-Castañeda: Appl. Opt. **33**, 5925 (1994)
7.25 J. Westerholm, J. Turunen, J. Huttunen: J. Opt. Soc. Am. A **11**, 1283 (1994)
7.26 J. Turunen: Pure Appl. Opt. **2**, 243 (1993)
7.27 M.T. Flores-Arias, C.R. Fernandez-Pousa, M.V. Perez, C. Bao, C. Gomez-Reino: J. Opt. Soc. Am. A **17**, 1007 (2000)
7.28 C. Gomez-Reino, M.T. Flores-Arias, M.V. Perez, C. Bao: Opt. Commun. **183**, 365 (2000)
7.29 M. Testorf, J. Jahns, N.A. Khilo, A.M. Goncharenko: Opt. Commun. **129**, 167 (1996)
7.30 A.W. Lohmann: *Optical Information Processing* (Universität Erlangen-Nijrnberg 1978), pp. 107–108

Chapter 8

8.1 D.A. Atchison, G.Smith: *Optics of the Human Eye* (Butterworth-Heinemann, Melbourne 2000), Sect. 1
8.2 S.G. ElHage, N.E. Leach: I. Cont. Lens Clin. **26**, 39 (1999)
8.3 N. Brown: Exp. Eye Res. **15**, 441 (1973)
8.4 H.J. Wyatt, R.F. Fisher: Eye **9**, 772 (1995)
8.5 L.F. Garner, M.K.H. Yap: Ophthal. Physiolog. Opt. **17**, 12 (1997)
8.6 L.F. Garner, G.S. Smith: Optom. Vision. Sci. **74**, 114 (1997)
8.7 B. Pierscionek: J. Biomed. Opt. **1**, 147 (1996)
8.8 A. Glasser, M.C.W. Campbell: Vision Res. **38**, 209 (1998)
8.9 S. Nakao, T. Ono, R. Nagata, K. Iwata: Jap. J. Clin. Ophthalmol. **23**, 903 (1969)
8.10 D.A. Palmer, J. G. Sivak: J. Opt. Soc. Am. **71**, 780 (1981)
8.11 J.G. Sivak: Am. J. Optom. Physiolog. Opt. **62**, 299 (1985)
8.12 M.C.W. Campbell: Vision Res. **24**, 409 (1984)
8.13 O. Pomerantzeff, M. Pankratov, G.J. Wang, P. Dufault: Am. J. Optom. Physiol. Opt. **61**, 166 (1984)
8.14 T. Raasch, V. Lakshminarayanan: Ophthal. Physiolog. Opt. **9**, 61 (1989)
8.15 I.H. Al-Ahdali, M.A. El-Messiery: Appl. Opt. **34**, 5738 (1995)
8.16 D.A. Atchison, G. Smith: Vision Res. **35**, 2529 (1995)
8.17 G. Smith, B.K. Pierscionek, D.A. Atchison: Ophthal. Physiolog. Opt. **11**, 359 (1991)
8.18 W.S. Jagger: Vision Res. **30**, 723 (1990)
8.19 G. Smith, D.A. Atchison, B.K. Pierscionek: J. Opt. Soc. Am. A **9**, 2111 (1992)
8.20 D.T. Moore: J. Opt. Soc. Am. **61**, 886 (1971)
8.21 G. Smith: J. Opt. Soc. Am. A **9**, 331 (1992)

8.22 J.W. Blaker: J. Opt. Soc. Am. **70**, 220 (1980)
8.23 M. Melgosa, A.J. Poza, E. Hita, C. Gomez-Reino: Atti Fond. G. Ronchi **50**, 895 (1995)
8.24 A. Gullstrand: `Appendix II. Optical Imagery´. In *Handbuch der Physiologischen Optik*, H. von Helmholtz 3rd edn. (Leopold Voss, Hamburg 1909); English translation, *Helmholtz's Treatise on Physiological Optics*, J.P.C. Southall, ed. (Optical Society of America, Washington 1924), Vol. 1
8.25 B.K. Pierscionek, D.Y.C. Chan: Optom. Vision Sci. **66**, 822 (1989)
8.26 G. Smith, D.A. Atchison: J. Opt. Soc. Am. A **14**, 2537 (1997)
8.27 L.F. Garner, S.O. Chuan, G. Smith: Clin. Exp. Optom. **81**, 145 (1998)

Chapter 9

9.1 J.C. Palais: Appl. Opt. **19**, 2011 (1980)
9.2 T. Sakamoto: Appl. Opt. **31**, 5184 (1992)
9.3 R.W. Gilsdorf, J.C. Palais: Appl. Opt. **33**, 3440 (1994)
9.4 S. Yuan, N.A. Riza: Appl. Opt. **38**, 3214 (1999)
9.5 H.R.D. Sunak: IEEE Trans. Instrum. Meas. IM-**27**, 557 (1989)
9.6 W.L. Emkey, C.A. Jack: J. Lightwave Technol. LT-**5**, 1156 (1987)
9.7 See, e.g., A.W. Snyder, J.D. Love: *Optical Waveguide Theory* (Chapman and Hall, London 1983), Chap. 15
9.8 S. Rios, R. Srivastava, C. Gomez-Reino: Opt. Commun. **119**, 517 (1995)
9.9 K. Thyagarajan, S.N. Sarkar, B.P. Pal: J. Lightwave Technol. LT-**5**, 1041 (1987)
9.10 J. Liñares, C. Gomez-Reino: Appl. Opt. **29**, 4003 (1990)
9.11 C. Gomez-Reino: Int. J. Optoelectron. **7**, 607 (1992)
9.12 J.R. Leger, G.J. Swanson, W.B. Weldkamp: Appl. Opt. **26**, 3439 (1987)
9.13 A. Aharoni, J.W. Goodman, Y. Amitai: Opt. Lett. **17**, 1310 (1992)
9.14 T.H. Zachos, J.C. Dyment: IEEE J. Quantum Electron. QE-**6**, 317 (1970)
9.15 E. Acosta, R.M. Gonzalez, C. Gomez-Reino: Opt. Lett. **16**, 627 (1991)
9.16 J.M. Stagaman, D.T. Moore: Appl. Opt. **23**, 1730 (1984)
9.17 C. Bao, C. Gomez-Reino, M.V. Perez: Opt. Commun. **172**, 1 (1999)
9.18 C. Gomez-Reino, E. Larrea, M.V. Perez, J.M. Cuadrado: Appl. Opt. **24**, 4379 (1985)
9.19 W.J. Tomlinson: Appl. Opt. **19**, 1127 (1980)
9.20 See, e.g., *Selfoc Product Guide* (NSG America, Inc., New Jersey 1997)
9.21 K. Hamanaka: Opt. Lett. **16**, 1222 (1991)
9.22 K. Hamanaka: Jpn. J. Appl. Phys. **31**, 1656 (1992)
9.23 T. Sakano, T. Matsumoto, K. Noguchi: Appl. Opt. **34**, 1815 (1995)
9.24 See, e.g., J.M. Lopez-Higuera, ed.: *Optical Sensors* (Cantabria University Press, Santander 1998)
9.25 B. Culshaw: Radio Electron. Eng. **52**, 283 (1982)
9.26 C.M. Davis: Opt. Eng. **24**, 347 (1985)
9.27 P.J. Murphy, T.P. Coursolle: Appl. Opt. **29**, 544 (1990)
9.28 W.H. Fu, C.F. Wang, W.Z. Tang, W. Zhou, J.H. Zhou: Int. J. Optoelectron. **7**, 121 (1992)
9.29 K. Uehira, K. Komiya: Appl. Opt. **29**, 4081 (1990)

Index

MIX
Papier aus verantwortungsvollen Quellen
Paper from responsible sources
FSC® C105338

If you have any concerns about our products,
you can contact us on
ProductSafety@springernature.com

In case Publisher is established outside the EU,
the EU authorized representative is:
Springer Nature Customer Service Center GmbH
Europaplatz 3, 69115 Heidelberg, Germany

Printed by Libri Plureos GmbH
in Hamburg, Germany